Cambridge Tracts in Mathematics
and Mathematical Physics

No. 42

IDEAL THEORY

IDEAL THEORY

BY

D. G. NORTHCOTT

Town Trust Professor of Pure Mathematics, University of Sheffield
Formerly Fellow of St John's College, Cambridge

CAMBRIDGE
AT THE UNIVERSITY PRESS
1953

PUBLISHED BY THE PRESS SYNDICATE OF THE UNIVERSITY OF CAMBRIDGE
The Pitt Building, Trumpington Street, Cambridge, United Kingdom

CAMBRIDGE UNIVERSITY PRESS
The Edinburgh Building, Cambridge CB2 2RU, UK
40 West 20th Street, New York NY 10011–4211, USA
477 Williamstown Road, Port Melbourne, VIC 3207, Australia
Ruiz de Alarcón 13, 28014 Madrid, Spain
Dock House, The Waterfront, Cape Town 8001, South Africa

http://www.cambridge.org

First published 1953
Reprinted 1960 1963 1965 1968 1972
First paperback edition 2004

A catalogue record for this book is available from the British Library

ISBN 0 521 05840 6 hardback
ISBN 0 521 60483 4 paperback

CONTENTS

Chapter III. SOME FUNDAMENTAL PROPERTIES OF NOETHERIAN RINGS

Chapter IV. THE ALGEBRAIC THEORY OF LOCAL RINGS

Chapter V. THE ANALYTIC THEORY OF LOCAL RINGS

PREFACE

The theory of ideals, in its modern form, is a contemporary development of mathematical knowledge to which mathematicians of to-day may justly point with pride. It not only has the generality and purity of logical structure, which is typical of so much of the work that has been done in recent years, but also it has contributed, in a substantial way, to the growth of an older branch of the mathematical tree, namely, algebraic geometry. It is not possible, in a volume of this size, both to give a useful account of the purely algebraic parts of our subject and also to give examples of the deeper applications, but it has proved possible to weave into a connected algebraic theory those results which play outstandingly important roles in the geometric applications. This is precisely what has been done. It is the author's hope that this tract will extend the interest taken in a new mathematical territory, by enabling the reader to travel, in relative comfort, along the road which pioneers like E. Noether, W. Krull, C. Chevalley and I. S. Cohen have constructed. Before he sets out, however, either to see the sights or with the intention of joining in the work when he reaches his destination, the traveller may fairly ask whether or not his present equipment will be sufficient for the proposed journey. Provided he has reached the standard of a good honours degree in mathematics he has no cause to worry. It is not necessary that he should have ever read a book or attended lectures on modern algebra, for in this respect the account is self-contained.

The notes, which follow Chapter V, will give the reader a general idea of the historical development of our subject, but his picture of the way in which the theory has grown will be distorted unless he remembers that the topics discussed form only a part of what might have been included under the broad heading 'Ideal Theory'. For this reason, he will find no mention of the work of certain mathematicians whose results would inevitably occupy prominent and important positions in a comprehensive treatise. There is another respect in which the notes would be misleading without some further comment. They indicate, in detail, the sources from which the materials for the tract have been collected. What they do not reveal, however, is the great personal debt the author owes to Professor E. Artin of Princeton University, who, during the years 1946–8, introduced him to the theory of ideals and developed his interest in it. More recently, the well-known writings of O. Zariski on algebraic geometry have been responsible for arousing, in the author, a particular interest in the properties of local rings, and he ventures to hope that this little volume will assist the mathematician, who has not had an intensive training in modern algebra, in reading Professor Zariski's papers.

It is a great pleasure to acknowledge the help which was given by Professor W. V. D. Hodge, who read the whole manuscript. Both the general plan of this tract and a number of points of detail have been very much improved as a result of his suggestions. The decision to devote a substantial amount of space to developments made during the last decade was largely due to his influence.

I am also very grateful to Dr Christine M. Hamill for the assistance which she has given in correcting the proofs.

D. G. N.

CAMBRIDGE
1952

NOTE TO THE READER

The numbering of theorems, propositions and lemmas is begun afresh in each chapter. When a reference is made to a result, which has previously been established, only the number is quoted if the result in question is in the same chapter as that in which the reference is made. In all other cases, both the number of the result and the number of subsection, in which it occurs, are given.

IDEAL THEORY

PRELIMINARIES

The starting point of the theory which is developed in the following pages, is the definition of the algebraic system which is known as a ring. Roughly speaking, a ring is a set of objects which can be *added* and *multiplied*, and which may be manipulated, in so far as these two operations are concerned, more or less in the natural manner. We shall now define precisely what it is that the algebraists call a ring, and then we shall deduce those elementary consequences of the definition, which are used constantly in the handling of formulae.

Suppose that we have a set R of objects, which we shall refer to as the *elements* of R, and which we shall denote by the letters a, b, c, and so on. Suppose, further, that with each ordered pair a, b of elements of R, there are associated two elements of R, which are called the *sum* and the *product* of a and b, and which are denoted by $a+b$ and by ab respectively. The set R (together with the operations of addition and multiplication) is said to form a *ring*, whenever the following six conditions are satisfied:

(1) $a+b=b+a$ *for all* a *and* b.

(2) $(a+b)+c=a+(b+c)$ *for all* a, b, *and* c.

(3) *There is an element* Θ *such that* $a+\Theta=a$ *for all* a.

(4) *For each element* a *there exists at least one element* x *such that* $a+x=\Theta$.

(5) $a(bc)=(ab)c$ *for all* a, b, *and* c.

(6) $a(b+c)=ab+ac$ *and* $(b+c)a=ba+ca$ *for all* a, b, *and* c.

In (1), the expression 'for all a and b' means, of course, 'for all pairs of elements a and b belonging to R', and similar interpretations are intended in (2), (3), (5) and (6). For brevity, instead of writing 'a belongs to R' we shall often write $a \in R$, and, more generally, if S is a subset of R we shall write $a \in S$ if we wish to indicate that a belongs to S. We shall now deduce those elementary consequences of the definition of a ring, to which a reference has already been made.

(7) *There is only one element* Θ *with the property that* $a+\Theta=a$ *for all* a.

For assume that $a+\Theta=a+\Theta'=a$ for all a. Then, on the one hand, $\Theta'+\Theta=\Theta'$; and, on the other hand, using (1), we have $\Theta'+\Theta=\Theta+\Theta'$ which is equal to Θ by hypothesis. Thus $\Theta'=\Theta'+\Theta=\Theta$. The element Θ is called the *zero element* of the ring.

(8) *If* a *and* b *are given the equation* $a+x=b$ *has one and only one solution.*

For by (4) there exists $y\in R$ such that $a+y=\Theta$, and then, since

$$a+(y+b)=(a+y)+b=\Theta+b=b+\Theta=b,$$

it follows that $y+b$ is one solution. Again, supposing $a+x=b$, we have

$$y+b=y+(a+x)=(y+a)+x=(a+y)+x=\Theta+x=x+\Theta=x,$$

which shows that $y+b$ is the only solution.

As a particular case we notice that the equation $a+x=\Theta$ has a unique solution. This solution will be denoted by $-a$. From $\Theta+\Theta=\Theta$ we see that $-\Theta=\Theta$, and from $(-a)+a=a+(-a)=\Theta$ we see that $-(-a)=a$. Again, by (1), (2) and (3), we have $(a+b)+(-a)=b$, and if we now add $-b$ to both sides of this equation we find that $-(a+b)=(-a)+(-b)$.

(9) $\Theta a=a\Theta=\Theta$ *for all* a.

In fact, $\Theta a=(\Theta+\Theta)a=\Theta a+\Theta a$ and since we also have $\Theta a+\Theta=\Theta a$ it follows, by (8), that $\Theta a=\Theta$. We can prove that $a\Theta=\Theta$ in a similar way.

(10) $-(ab)=(-a)b=a(-b)$, *and* $(-a)(-b)=ab$.

To see this we note that

$$\Theta=\Theta b=(a+(-a))b=ab+(-a)b$$

which shows that $-(ab)=(-a)b$, and the proof that

$$-(ab)=a(-b)$$

is similar. If we now take the negatives of both sides of the equation $-(ab)=(-a)b$ we obtain

$$ab=-(-a)b=(-a)(-b).$$

It is usually convenient to write $a-b$ in place of $a+(-b)$. It follows immediately that $-(a-b)=b-a$; that $c(a-b)=ca-cb$; and that $(a-b)c=ac-bc$.

To the above observations, we add the remark that from (2) and (5) it follows, in the usual way, that we can use symbols such as $a_1+a_2+\ldots+a_n$ and $a_1a_2\ldots a_n$, that is, sums and products of more than two terms, without ambiguity.

Commutative rings. The ring R is called *commutative* if $ab=ba$ for all a and b. In such a ring, permuting the terms of a product $a_1a_2\ldots a_n$ does not change its value. Roughly speaking, in a commutative ring all the usual algebraic manipulations are permissible, except those which involve cancellation or division. The rings, which will concern us, are such that, usually, we cannot conclude from $ab=ac$ and $a\neq\Theta$ that $b=c$; and it will hardly ever be true that $a\neq\Theta$ implies that the equation $ax=b$ has a solution.

Rings with a unit element. A ring R, commutative or not, is said to have an element e as a *unit element* if $e\neq\Theta$, and if $ea=ae=a$ for all a. A ring has at most one unit element, for if

$$ea=ae=e'a=ae'=a$$

for all a, then ee' is equal both to e and to e', hence e and e' are the same.

It is extremely easy to give a large number of different examples of a ring, but we shall not stop to try and indicate the scope of this concept. However, it is a good idea for the reader to keep a definite example in mind, and for this the polynomials in n variables, with complex numbers as coefficients, will serve excellently. Such polynomials are easily seen to form a ring, and it was the study of this ring (in connexion with algebraic geometry) which gave rise to a large part of contemporary ideal theory.

CHAPTER I

THE PRIMARY DECOMPOSITION

1·1. A convention. Now that we have given a formal definition of a ring, we can begin the systematic development of our subject. The rings that we shall consider will all be commutative, and they will all have a unit element. It is therefore convenient to use the word 'ring' in a more restricted sense than is customary in modern algebra, and for this reason we lay down the following convention: *From now on 'ring' will always mean a commutative ring with a unit element.* The zero element and the unit element of a ring R will be denoted by 0 and 1 respectively, or, if we are concerned with several rings at the same time, by 0_R and 1_R.

1·2. Ideals and their calculus. Let R be a ring (commutative and with a unit element), and let $\mathfrak{a}$ be a non-empty subset of R, then $\mathfrak{a}$ is called an *ideal* of R in all cases where the following two conditions are satisfied:

(1) *Whenever a_1 and a_2 belong to $\mathfrak{a}$, then $a_1 \pm a_2$ both belong to $\mathfrak{a}$.*

(2) *If $a \in \mathfrak{a}$, then $ra \in \mathfrak{a}$ for all $r \in R$.*

A trivial example of an ideal is obtained by taking $\mathfrak{a}$ to be the whole ring. We shall call an ideal which is not the whole ring a *proper ideal*; for example, the set consisting only of the zero element is not only an ideal (this follows immediately from the definition), but it is also a proper ideal, for by the very definition of the unit element, 1 and 0 are different. Let us note that every ideal $\mathfrak{a}$ contains the zero element, for we can choose $a \in \mathfrak{a}$ (since $\mathfrak{a}$ is not empty) and then $0a = 0$ will belong to $\mathfrak{a}$ by (2); further, since $-a = 0 - a$, it follows from (1) that if $a \in \mathfrak{a}$ then $-a \in \mathfrak{a}$.

As far as is convenient we shall use small German letters $\mathfrak{a}$, $\mathfrak{b}$, $\mathfrak{c}$, etc., to denote ideals, and we shall employ small Latin and Greek letters to denote elements.

The four basic ways of combining ideals are known as *addition*, *multiplication*, *intersection*, and *residual division*.

Addition. Suppose that $\mathfrak{a}$ and $\mathfrak{b}$ are two given ideals, and that $\mathfrak{c}$ is the set of all elements which can be written in the form $a+b$, where $a \in \mathfrak{a}$ and $b \in \mathfrak{b}$. Then $\mathfrak{c}$ is an ideal. To see this suppose that $x_1 \in \mathfrak{c}$, that $x_2 \in \mathfrak{c}$, and that $r \in R$, then $x_1 = a_1 + b_1$, $x_2 = a_2 + b_2$ where a_1, a_2 belong to $\mathfrak{a}$ and where b_1, b_2 belong to $\mathfrak{b}$, and we also have

$$x_1 + x_2 = (a_1 + a_2) + (b_1 + b_2), \quad x_1 - x_2 = (a_1 - a_2) + (b_1 - b_2),$$
$$rx_1 = (ra_1) + (rb_1).$$

But $\mathfrak{a}$ and $\mathfrak{b}$ are ideals so $a_1 + a_2$, $a_1 - a_2$, ra_1 are all in $\mathfrak{a}$, and $b_1 + b_2$, $b_1 - b_2$, rb_1 are all in $\mathfrak{b}$. This shows that $x_1 + x_2$, $x_1 - x_2$, and rx_1 are all in $\mathfrak{c}$, and thereby establishes that $\mathfrak{c}$ is an ideal. This ideal is called the *sum* of $\mathfrak{a}$ and $\mathfrak{b}$ and is denoted by $\mathfrak{a} + \mathfrak{b}$. It is clear that $\mathfrak{a} + \mathfrak{b} = \mathfrak{b} + \mathfrak{a}$. Further, if $\mathfrak{a}_1$, $\mathfrak{a}_2$, $\mathfrak{a}_3$ are any three ideals then $\mathfrak{a}_1 + (\mathfrak{a}_2 + \mathfrak{a}_3) = (\mathfrak{a}_1 + \mathfrak{a}_2) + \mathfrak{a}_3$, for both $\mathfrak{a}_1 + (\mathfrak{a}_2 + \mathfrak{a}_3)$ and $(\mathfrak{a}_1 + \mathfrak{a}_2) + \mathfrak{a}_3$ consist of all elements of the form $a_1 + a_2 + a_3$, where $a_i \in \mathfrak{a}_i$ for $i = 1, 2, 3$. Now every element of $\mathfrak{a}$ can be written in the form $a + 0$ where $a \in \mathfrak{a}$, which (since $0 \in \mathfrak{b}$) shows that $\mathfrak{a}$ is contained in $\mathfrak{a} + \mathfrak{b}$. Quite generally, if A and B are two subsets of R, we write $A \subseteq B$ or $B \supseteq A$ whenever every element of A is an element of B (i.e. whenever A is contained in B), and we write $A \subset B$ or $B \supset A$ if $A \subseteq B$ and there is at least one element of B not contained in A (i.e. whenever the inclusion is strict). We have therefore proved that $\mathfrak{a} \subseteq \mathfrak{a} + \mathfrak{b}$, and since $\mathfrak{a} + \mathfrak{b} = \mathfrak{b} + \mathfrak{a}$, it follows also that $\mathfrak{b} \subseteq \mathfrak{a} + \mathfrak{b}$.

Multiplication. This time, supposing that $\mathfrak{a}$ and $\mathfrak{b}$ are given ideals, we let $\mathfrak{c}$ consist of all elements which can be written as a finite sum of products ab, where $a \in \mathfrak{a}$ and $b \in \mathfrak{b}$. $\mathfrak{c}$ is an ideal. For suppose that $x_1, x_2 \in \mathfrak{c}$ and that $r \in R$, then

$$x_1 = a_1 b_1 + a_2 b_2 + \ldots + a_p b_p, \quad x_2 = a_1' b_1' + a_2' b_2' + \ldots + a_q' b_q',$$

where the a_i and the a_j' are in $\mathfrak{a}$, and the b_i and b_j' are in $\mathfrak{b}$. From this we obtain

$$x_1 + x_2 = a_1 b_1 + \ldots + a_q' b_q',$$

$$x_1 - x_2 = a_1 b_1 + \ldots + a_p b_p + (-a_1') b_1' + \ldots + (-a_q') b_q',$$

and
$$rx_1 = (ra_1) b_1 + \ldots + (ra_p) b_p.$$

Now $-a_j' \in \mathfrak{a}$ and $ra_i \in \mathfrak{a}$, which establishes that $x_1 + x_2$, $x_1 - x_2$, and rx_1 are all in $\mathfrak{c}$, and, consequently, that $\mathfrak{c}$ is an ideal. The ideal that we have just constructed is called the *product* of $\mathfrak{a}$ and $\mathfrak{b}$, and is denoted by $\mathfrak{ab}$. We see at once that $\mathfrak{ab}=\mathfrak{ba}$, and, if $\mathfrak{a}_1, \mathfrak{a}_2, \mathfrak{a}_3$ are any three ideals, that $\mathfrak{a}_1(\mathfrak{a}_2\mathfrak{a}_3)=(\mathfrak{a}_1\mathfrak{a}_2)\mathfrak{a}_3$, for both sides of the latter equation consist of all elements which can be written as a finite sum of products $a_1a_2a_3$, where $a_i \in \mathfrak{a}_i$ for $i=1,2,3$. Further,

$$\mathfrak{ab}\subseteq\mathfrak{a}, \quad \mathfrak{ab}\subseteq\mathfrak{b} \quad \text{and} \quad \mathfrak{a}(\mathfrak{b}_1+\mathfrak{b}_2)=\mathfrak{ab}_1+\mathfrak{ab}_2.$$

Here the first two assertions are quite obvious; let us prove the third. Since $\mathfrak{b}_1\subseteq\mathfrak{b}_1+\mathfrak{b}_2$ we have $\mathfrak{ab}_1\subseteq\mathfrak{a}(\mathfrak{b}_1+\mathfrak{b}_2)$ and similarly $\mathfrak{ab}_2\subseteq\mathfrak{a}(\mathfrak{b}_1+\mathfrak{b}_2)$; consequently

$$\mathfrak{ab}_1+\mathfrak{ab}_2\subseteq\mathfrak{a}(\mathfrak{b}_1+\mathfrak{b}_2)+\mathfrak{a}(\mathfrak{b}_1+\mathfrak{b}_2)=\mathfrak{a}(\mathfrak{b}_1+\mathfrak{b}_2),$$

for a moment's reflexion shows that an ideal is unaltered by adding it to itself. Again, if $a\in\mathfrak{a}$, $b_1\in\mathfrak{b}_1$, and $b_2\in\mathfrak{b}_2$, then

$$a(b_1+b_2)=ab_1+ab_2\in(\mathfrak{ab}_1+\mathfrak{ab}_2),$$

and therefore any sum of terms such as $a(b_1+b_2)$ will also belong to $\mathfrak{ab}_1+\mathfrak{ab}_2$. In other words, every element of $\mathfrak{a}(\mathfrak{b}_1+\mathfrak{b}_2)$ belongs to $\mathfrak{ab}_1+\mathfrak{ab}_2$, i.e. $\mathfrak{a}(\mathfrak{b}_1+\mathfrak{b}_2)\subseteq\mathfrak{ab}_1+\mathfrak{ab}_2$. Combining this with $\mathfrak{ab}_1+\mathfrak{ab}_2\subseteq\mathfrak{a}(\mathfrak{b}_1+\mathfrak{b}_2)$ we obtain $\mathfrak{ab}_1+\mathfrak{ab}_2=\mathfrak{a}(\mathfrak{b}_1+\mathfrak{b}_2)$.

Intersection. Let $\mathfrak{a}_i$ $(i\in I)$ be a finite or infinite set of ideals, the range I of the suffix i being quite arbitrary. The elements which belong to all the $\mathfrak{a}_i$ form a set which is called their *intersection*, and which is denoted by $\bigcap_{i\in I}\mathfrak{a}_i$, or, more casually, by $\bigcap\mathfrak{a}_i$. This intersection is not empty, for all the $\mathfrak{a}_i$ contain the element 0, and a trivial verification shows that it is in fact an ideal. The intersection of a finite set of ideals $\mathfrak{a}_1, \mathfrak{a}_2, \ldots, \mathfrak{a}_n$ we write either as $\bigcap_{i=1}^{n}\mathfrak{a}_i$, or as $\mathfrak{a}_1\cap\mathfrak{a}_2\cap\ldots\cap\mathfrak{a}_n$. We proved earlier that $\mathfrak{ab}\subseteq\mathfrak{a}$ and $\mathfrak{ab}\subseteq\mathfrak{b}$. This can now be written more conveniently as $\mathfrak{ab}\subseteq\mathfrak{a}\cap\mathfrak{b}$.

Residual division. Once again suppose that $\mathfrak{a}$ and $\mathfrak{b}$ are two ideals, and let us denote by $\mathfrak{c}$ the set of all elements x such that $xb\in\mathfrak{a}$ for all $b\in\mathfrak{b}$. $\mathfrak{c}$ is an ideal. For suppose that $x_1, x_2\in\mathfrak{c}$ and that $r\in R$, then for any $b\in\mathfrak{b}$ we have

$$(x_1+x_2)b=x_1b+x_2b,$$

$$(x_1-x_2)b=x_1b-x_2b \quad \text{and} \quad (rx_1)b=r(x_1b).$$

Now x_1b and x_2b belong to $\mathfrak{a}$ by the definition of $\mathfrak{c}$, so x_1b+x_2b and x_1b-x_2b belong to $\mathfrak{a}$; also as $x_1b\in\mathfrak{a}$ we have $r(x_1b)\in\mathfrak{a}$. This proves that $x_1 \pm x_2$ and rx_1 are all in $\mathfrak{c}$, which shows that $\mathfrak{c}$ is an ideal. $\mathfrak{c}$ is known as the *residual quotient* of $\mathfrak{a}$ and $\mathfrak{b}$, and is denoted by $\mathfrak{a} : \mathfrak{b}$. Since from $x \in (\mathfrak{a} : \mathfrak{b})$ and $b \in \mathfrak{b}$ follows $xb \in \mathfrak{a}$, we see that $(\mathfrak{a} : \mathfrak{b})\,\mathfrak{b} \subseteq \mathfrak{a}$; in fact $\mathfrak{a} : \mathfrak{b}$ is the largest ideal which when multiplied by $\mathfrak{b}$ yields an ideal contained in $\mathfrak{a}$. We note too that if $a \in \mathfrak{a}$ then certainly $ab \in \mathfrak{a}$ for all $b \in \mathfrak{b}$, which gives us the relation $\mathfrak{a} \subseteq (\mathfrak{a} : \mathfrak{b})$.

For convenience we collect together the basic formulae of our calculus, which have now been established, and add to them some new ones of a more advanced character.

Proposition 1.

(1) $\mathfrak{a}+\mathfrak{b}=\mathfrak{b}+\mathfrak{a}$; $\mathfrak{a}+(\mathfrak{b}+\mathfrak{c})=(\mathfrak{a}+\mathfrak{b})+\mathfrak{c}$.

(2) $\mathfrak{a}\mathfrak{b}=\mathfrak{b}\mathfrak{a}$; $\mathfrak{a}(\mathfrak{b}\mathfrak{c})=(\mathfrak{a}\mathfrak{b})\,\mathfrak{c}$; $\mathfrak{a}(\mathfrak{b}+\mathfrak{c})=\mathfrak{a}\mathfrak{b}+\mathfrak{a}\mathfrak{c}$.

(3) $\mathfrak{a} \subseteq \mathfrak{a}+\mathfrak{b}$; $\mathfrak{a}\mathfrak{b} \subseteq (\mathfrak{a} \cap \mathfrak{b})$.

(4) $(\mathfrak{a}:\mathfrak{b})\,\mathfrak{b} \subseteq \mathfrak{a}$; $\mathfrak{a} \subseteq (\mathfrak{a}:\mathfrak{b})$.

(5) $(\bigcap \mathfrak{a}_i):\mathfrak{b} = \bigcap(\mathfrak{a}_i:\mathfrak{b})$.

(6) $(\mathfrak{a}:\mathfrak{b}):\mathfrak{c} = \mathfrak{a}:(\mathfrak{b}\mathfrak{c})$.

(7) $\mathfrak{a}:(\mathfrak{b}_1+\mathfrak{b}_2+\ldots+\mathfrak{b}_n) = (\mathfrak{a}:\mathfrak{b}_1) \cap (\mathfrak{a}:\mathfrak{b}_2) \cap \ldots \cap (\mathfrak{a}:\mathfrak{b}_n)$.

(8) $\mathfrak{a}:\mathfrak{b} = \mathfrak{a}:(\mathfrak{a}+\mathfrak{b})$.

Proof. (1), (2), (3) and (4) have already been established.

(5) Let $x \in (\bigcap \mathfrak{a}_i):\mathfrak{b}$ and let $b \in \mathfrak{b}$, then $xb \in \bigcap \mathfrak{a}_i$ so that $xb \in \mathfrak{a}_i$ for all i. Keep i fixed, then $xb \in \mathfrak{a}_i$ for all $b \in \mathfrak{b}$; i.e. $x \in (\mathfrak{a}_i : \mathfrak{b})$. We have now shown that $x \in (\mathfrak{a}_i:\mathfrak{b})$ for all i, consequently $x \in \bigcap (\mathfrak{a}_i:\mathfrak{b})$. Since x was any element of $(\bigcap \mathfrak{a}_i):\mathfrak{b}$, we have proved that $(\bigcap \mathfrak{a}_i):\mathfrak{b} \subseteq \bigcap(\mathfrak{a}_i:\mathfrak{b})$. Now suppose that $y \in \bigcap(\mathfrak{a}_i:\mathfrak{b})$ and let $b \in \mathfrak{b}$. For each i we have $y \in (\mathfrak{a}_i : \mathfrak{b})$, hence $yb \in \mathfrak{a}_i$, and therefore $yb \in \bigcap \mathfrak{a}_i$. But as b was any element of $\mathfrak{b}$, it follows from $yb \in \bigcap \mathfrak{a}_i$ that $y \in (\bigcap \mathfrak{a}_i):\mathfrak{b}$. Since y was an arbitrary element of $\bigcap(\mathfrak{a}_i:\mathfrak{b})$ this proves that $\bigcap(\mathfrak{a}_i:\mathfrak{b}) \subseteq (\bigcap \mathfrak{a}_i):\mathfrak{b}$. Combining this last relation with $(\bigcap \mathfrak{a}_i):\mathfrak{b} \subseteq \bigcap(\mathfrak{a}_i:\mathfrak{b})$ we obtain the required result.

(6) Let $x \in (\mathfrak{a}:\mathfrak{b}):\mathfrak{c}$ and let b_i, c_i belong to $\mathfrak{b}$ and $\mathfrak{c}$ respectively for $1 \leqslant i \leqslant s$. Then $xc_i \in (\mathfrak{a}:\mathfrak{b})$ so that $xb_ic_i \in \mathfrak{a}$. If we now sum over i we obtain $x(b_1c_1+b_2c_2+\ldots+b_sc_s) \in \mathfrak{a}$. But $b_1c_1+b_2c_2+\ldots+b_sc_s$ can be any element of $\mathfrak{b}\mathfrak{c}$, hence we have proved that $x \in \mathfrak{a}:(\mathfrak{b}\mathfrak{c})$. Now let $y \in \mathfrak{a}:(\mathfrak{b}\mathfrak{c})$, let $b \in \mathfrak{b}$ and let $c \in \mathfrak{c}$. Then $ybc \in \mathfrak{a}$. Since this

holds for all $b \in \mathfrak{b}$, we must have $yc \in (\mathfrak{a} : \mathfrak{b})$, and as $yc \in (\mathfrak{a} : \mathfrak{b})$ for all $c \in \mathfrak{c}$, this proves that $y \in (\mathfrak{a} : \mathfrak{b}) : \mathfrak{c}$. Our combined results tell us that $(\mathfrak{a} : \mathfrak{b}) : \mathfrak{c} \subseteq \mathfrak{a} : (\mathfrak{b}\mathfrak{c})$, and also that $\mathfrak{a} : (\mathfrak{b}\mathfrak{c}) \subseteq (\mathfrak{a} : \mathfrak{b}) : \mathfrak{c}$. This is equivalent to what we had to prove.

(7) Suppose first that $n = 2$. Let $x \in \mathfrak{a} : (\mathfrak{b}_1 + \mathfrak{b}_2)$, let $b_1 \in \mathfrak{b}_1$, and let $b_2 \in \mathfrak{b}_2$. Then $b_1 + 0 \in (\mathfrak{b}_1 + \mathfrak{b}_2)$ so that $xb_1 = x(b_1 + 0) \in \mathfrak{a}$. This holds for all $b_1 \in \mathfrak{b}_1$, consequently $x \in \mathfrak{a} : \mathfrak{b}_1$. Similarly, $x \in \mathfrak{a} : \mathfrak{b}_2$, and therefore $x \in (\mathfrak{a} : \mathfrak{b}_1) \cap (\mathfrak{a} : \mathfrak{b}_2)$. Now assume that $y \in (\mathfrak{a} : \mathfrak{b}_1) \cap (\mathfrak{a} : \mathfrak{b}_2)$, let $b_1 \in \mathfrak{b}_1$, and let $b_2 \in \mathfrak{b}_2$. Since $y \in (\mathfrak{a} : \mathfrak{b}_1)$ we have $yb_1 \in \mathfrak{a}$, and since $y \in (\mathfrak{a} : \mathfrak{b}_2)$ we have $yb_2 \in \mathfrak{a}$. By addition we obtain $y(b_1 + b_2) \in \mathfrak{a}$, but as $b_1 + b_2$ can be any element of $\mathfrak{b}_1 + \mathfrak{b}_2$ it follows that $y \in \mathfrak{a} : (\mathfrak{b}_1 + \mathfrak{b}_2)$. We have now established

$$\mathfrak{a} : (\mathfrak{b}_1 + \mathfrak{b}_2) \subseteq (\mathfrak{a} : \mathfrak{b}_1) \cap (\mathfrak{a} : \mathfrak{b}_2) \quad \text{and} \quad (\mathfrak{a} : \mathfrak{b}_1) \cap (\mathfrak{a} : \mathfrak{b}_2) \subseteq \mathfrak{a} : (\mathfrak{b}_1 + \mathfrak{b}_2),$$

which is equivalent to (7) when $n = 2$. The extension to a general n is now obtained by a simple induction. It should be noted that in writing $\mathfrak{b}_1 + \mathfrak{b}_2 + \ldots + \mathfrak{b}_n$ we are already making use of the associative law of addition, namely, $\mathfrak{a} + (\mathfrak{b} + \mathfrak{c}) = (\mathfrak{a} + \mathfrak{b}) + \mathfrak{c}$.

(8) By (7) $\mathfrak{a} : (\mathfrak{a} + \mathfrak{b}) = (\mathfrak{a} : \mathfrak{a}) \cap (\mathfrak{a} : \mathfrak{b})$. It is clear that $\mathfrak{a} : \mathfrak{a}$ is the whole ring R, accordingly $\mathfrak{a} : (\mathfrak{a} + \mathfrak{b}) = R \cap (\mathfrak{a} : \mathfrak{b}) = \mathfrak{a} : \mathfrak{b}$.

Remarks. The relation (5) is particularly important, for it says that an arbitrary intersection may be divided term by term.

We have already had occasion to note that an expression such as $\mathfrak{b}_1 + \mathfrak{b}_2 + \ldots + \mathfrak{b}_n$ is effectively unambiguous on account of the associative law of addition. A similar observation applies to a product $\mathfrak{b}_1 \mathfrak{b}_2 \ldots \mathfrak{b}_n$.

1·3. The ideal generated by a set. Let A be an arbitrary non-empty set of elements of our ring R. The aggregate of all elements which can be written in the form $\Sigma r_i a_i$, where $r_i \in R$, where $a_i \in A$, and where the number of terms in the sum is finite, is an ideal. The verification is extremely simple and will be left to the reader. This ideal, which is known as the *ideal generated by* A, contains every element of A, for if $a \in A$ then $1a = a$ belongs to the ideal in question. Further, every ideal which contains A will also contain the ideal generated by A, so that the ideal generated by A may be characterized as the smallest ideal containing A. As examples we note that the sum of two ideals

$\mathfrak{a}$ and $\mathfrak{b}$ is generated by the union of $\mathfrak{a}$ and $\mathfrak{b}$, that is, by the set obtained by taking all the elements of $\mathfrak{a}$ and all the elements of $\mathfrak{b}$; while $\mathfrak{a}\mathfrak{b}$ is generated by the set of all products ab where $a \in \mathfrak{a}$ and $b \in \mathfrak{b}$.

If A consists of a finite number of elements, say $a_1, a_2, \ldots, a_n$, then the ideal which they generate is denoted by $(a_1, a_2, \ldots, a_n)$ and it consists of all elements which can be written in the form $r_1 a_1 + r_2 a_2 + \ldots + r_n a_n$, where the r_i may be any elements of R. Such an ideal is said to be *finitely generated*, and the elements a_i are called a *base* or *basis* of the ideal. We note that

$$(a_1, a_2, \ldots, a_m) + (b_1, b_2, \ldots, b_n) = (a_1, \ldots, a_m, b_1, \ldots, b_n),$$

and that $$(a_1, a_2, \ldots, a_m)(b_1, b_2, \ldots, b_n) = (\ldots, a_i b_j, \ldots)$$

where the base on the right-hand side consists of all products $a_i b_j$.

An ideal (a) generated by a single element is known as a *principal ideal*.

1·4. Prime ideals. An ideal $\mathfrak{p}$ is called a *prime ideal* if whenever $ab \in \mathfrak{p}$ at least one of a and b belongs to $\mathfrak{p}$. Expressed in another way, our definition states that $\mathfrak{p}$ is prime if, and only if, from $ab \in \mathfrak{p}$ and $a \notin \mathfrak{p}$ always follows $b \in \mathfrak{p}$. Here we have made use of the symbol $\notin$ (cancelled epsilon) for the first time. It stands for the phrase 'does not belong to'.

PROPOSITION 2. *Let $\mathfrak{p}$ be a prime ideal, and suppose that $a_1 a_2 \ldots a_n \in \mathfrak{p}$, then for at least one value of i we have $a_i \in \mathfrak{p}$. Further, if $\mathfrak{a}_1, \mathfrak{a}_2, \ldots, \mathfrak{a}_n$ are ideals and $\mathfrak{a}_1 \mathfrak{a}_2 \ldots \mathfrak{a}_n \subseteq \mathfrak{p}$, then $\mathfrak{a}_i \subseteq \mathfrak{p}$ for at least one value of i.*

Proof. Suppose that $a_1 a_2 \ldots a_n \in \mathfrak{p}$ and that no a_i belongs to $\mathfrak{p}$. We shall obtain a contradiction. We have $a_1(a_2 a_3 \ldots a_n) \in \mathfrak{p}$ and $a_1 \notin \mathfrak{p}$, hence, by the definition of a prime ideal, $a_2 a_3 \ldots a_n \in \mathfrak{p}$. We now repeat the argument and obtain in succession, $a_3 \ldots a_n \in \mathfrak{p}$, $a_4 \ldots a_n \in \mathfrak{p}$, and finally, $a_n \in \mathfrak{p}$. This is the required contradiction. Next assume that $\mathfrak{a}_1 \mathfrak{a}_2 \ldots \mathfrak{a}_n \subseteq \mathfrak{p}$, but that no $\mathfrak{a}_i$ is contained in $\mathfrak{p}$. For each i we can choose $a_i \in \mathfrak{a}_i$ so that $a_i \notin \mathfrak{p}$, and then, by the first part, $a_1 a_2 \ldots a_n \notin \mathfrak{p}$. However, $a_1 a_2 \ldots a_n \in \mathfrak{a}_1 \mathfrak{a}_2 \ldots \mathfrak{a}_n$ and therefore *a fortiori* $a_1 a_2 \ldots a_n \in \mathfrak{p}$, which is again a contradiction.

1·5. Primary ideals. An ideal $\mathfrak{q}$ is called a *primary ideal* if the conditions $ab \in \mathfrak{q}$ and $a \notin \mathfrak{q}$ always imply that some positive power of b is in $\mathfrak{q}$. Let us note that prime ideals are always primary.

We can now indicate the lines on which the remainder of this chapter will be developed. The prime and primary ideals play roles which are (very roughly) similar to those played by prime numbers and by prime-power numbers in elementary arithmetic. They are ideals of a particularly simple type, and enjoy many special properties. Our first object will be to derive these properties, and then later we shall consider when and how a general ideal may be decomposed into primary components.

The proposition which follows shows that with every primary ideal there is associated a definite prime ideal.

PROPOSITION 3. *Let $\mathfrak{q}$ be a given primary ideal, and let $\mathfrak{p}$ denote the set of all elements x such that $x^n \in \mathfrak{q}$ for at least one positive integral value of n. Then $\mathfrak{p}$ is a prime ideal which contains $\mathfrak{q}$, and which is contained in every other prime ideal which contains $\mathfrak{q}$.*

Proof. First we show that $\mathfrak{p}$ is an ideal. Let $x, y \in \mathfrak{p}$ and let $r \in R$. Then there exist integers m and n, such that $x^m \in \mathfrak{q}$ and $y^n \in \mathfrak{q}$. Now $(x+y)^{m+n}$ can be written as a sum of terms $x^\mu y^\nu$, where $0 \leqslant \mu$, $0 \leqslant \nu$, and where $\mu + \nu = m + n$. Accordingly, we have either $\mu \geqslant m$ or $\nu \geqslant n$. In the first case $x^\mu \in \mathfrak{q}$, and in the second case $y^\nu \in \mathfrak{q}$, so that in either case $x^\mu y^\nu \in \mathfrak{q}$. This shows that $(x+y)^{m+n} \in \mathfrak{q}$, consequently, by the definition of $\mathfrak{p}$, $x+y \in \mathfrak{p}$. A similar argument shows that $x - y \in \mathfrak{p}$. Again, $(rx)^m = r^m x^m \in \mathfrak{q}$ and therefore $rx \in \mathfrak{p}$. Since $x+y$, $x-y$, and rx are all in $\mathfrak{p}$, $\mathfrak{p}$ is an ideal.

We shall now prove that $\mathfrak{p}$ is prime. Assume that $ab \in \mathfrak{p}$ and that $a \notin \mathfrak{p}$. It will be enough to show that $b \in \mathfrak{p}$. Since $ab \in \mathfrak{p}$ there is a positive integer s such that $a^s b^s \in \mathfrak{q}$. But $a^s \notin \mathfrak{q}$, for otherwise a would belong to $\mathfrak{p}$, consequently (since $\mathfrak{q}$ is primary) some power of b^s is in $\mathfrak{q}$. This is the same as saying that some power of b is in $\mathfrak{q}$, or that $b \in \mathfrak{p}$.

It is obvious from the definition that $\mathfrak{p} \supseteq \mathfrak{q}$. Let $\mathfrak{p}'$ be any prime ideal containing $\mathfrak{q}$ and let $x \in \mathfrak{p}$. Then, with a suitable integer m, $x^m \in \mathfrak{q} \subseteq \mathfrak{p}'$. Since $\mathfrak{p}'$ is prime it follows from Proposition 2 that $x \in \mathfrak{p}'$. Thus $\mathfrak{p} \subseteq \mathfrak{p}'$ and the proof is complete.

If $\mathfrak{q}$ is a primary ideal and if $\mathfrak{p}$ is the prime ideal of Proposition 3, we shall say that $\mathfrak{q}$ *belongs to* $\mathfrak{p}$, and also that $\mathfrak{q}$ *is* $\mathfrak{p}$*-primary*.

COROLLARY 1. *If* $\mathfrak{q}$ *is* $\mathfrak{p}$*-primary,* $ab \in \mathfrak{q}$, *and* $a \notin \mathfrak{p}$ *then* $b \in \mathfrak{q}$.

This follows immediately from the definitions. In the next two corollaries $\mathfrak{a}$ and $\mathfrak{b}$ denote ideals.

COROLLARY 2. *If* $\mathfrak{q}$ *is* $\mathfrak{p}$*-primary,* $\mathfrak{a}\mathfrak{b} \subseteq \mathfrak{q}$, *and* $\mathfrak{a} \nsubseteq \mathfrak{p}$ *then* $\mathfrak{b} \subseteq \mathfrak{q}$.

We can choose $a_0 \in \mathfrak{a}$ so that $a_0 \notin \mathfrak{p}$. If now b is an arbitrary element of $\mathfrak{b}$, we have $a_0 b \in \mathfrak{q}$ and $a_0 \notin \mathfrak{p}$, consequently (by Corollary 1) $b \in \mathfrak{q}$. This completes the proof.

An important consequence of Corollary 2 is

COROLLARY 3. *If* $\mathfrak{q}$ *is* $\mathfrak{p}$*-primary and if* $\mathfrak{a} \nsubseteq \mathfrak{p}$, *then* $\mathfrak{q} : \mathfrak{a} = \mathfrak{q}$.

Proof. By (4) of Proposition 1, $\mathfrak{a}(\mathfrak{q} : \mathfrak{a}) \subseteq \mathfrak{q}$, hence, by our hypotheses and by Corollary 2, $\mathfrak{q} : \mathfrak{a} \subseteq \mathfrak{q}$. Since, in any case, $\mathfrak{q} \subseteq \mathfrak{q} : \mathfrak{a}$, this proves our assertion.

The following lemma is of great use in deciding when the primary-prime relationship holds.

LEMMA 1. *Suppose that* $\mathfrak{p}'$ *and* $\mathfrak{q}'$ *are two ideals for which the following conditions are satisfied:*

(*a*) $\mathfrak{p}' \supseteq \mathfrak{q}'$.

(*b*) *If* $x \in \mathfrak{p}'$, *then some positive power of* x *is in* $\mathfrak{q}'$.

(*c*) *If* $ab \in \mathfrak{q}'$ *and* $a \notin \mathfrak{p}'$, *then* $b \in \mathfrak{q}'$.

In these circumstances $\mathfrak{p}'$ *is a prime ideal, and* $\mathfrak{q}'$ *is a primary deal belonging to* $\mathfrak{p}'$.

Proof. We begin by showing that $\mathfrak{q}'$ is primary. Assume that $ab \in \mathfrak{q}'$ and that $b \notin \mathfrak{q}'$, then, by (*c*), $a \in \mathfrak{p}'$. Further, by (*b*), $a^n \in \mathfrak{q}'$ for some positive integer n. This proves that $\mathfrak{q}'$ is primary; let it be $\mathfrak{p}$-primary. From (*b*), we see immediately that $\mathfrak{p}' \subseteq \mathfrak{p}$. Let $x \in \mathfrak{p}$, then if we show that $x \in \mathfrak{p}'$ we shall have proved that $\mathfrak{p} = \mathfrak{p}'$ and this will establish the lemma. Choose the smallest positive integer i such that $x^i \in \mathfrak{q}'$. On one hand, if $i = 1$ we have $x \in \mathfrak{q}' \subseteq \mathfrak{p}'$; on the other hand, if $i > 1$ then $x^i = xx^{i-1} \in \mathfrak{q}'$ and $x^{i-1} \notin \mathfrak{q}'$, consequently, using (*c*), we have $x \in \mathfrak{p}'$. Thus in either case we have shown that $x \in \mathfrak{p}'$, as required.

There are three further properties of prime and primary ideals that we shall need. These are established in Propositions 4, 5 and 6.

PROPOSITION 4. *If $\mathfrak{q}_1, \mathfrak{q}_2, \ldots, \mathfrak{q}_n$ are all of them $\mathfrak{p}$-primary ideals, then $\mathfrak{q} = \mathfrak{q}_1 \cap \mathfrak{q}_2 \cap \ldots \cap \mathfrak{q}_n$ is also a $\mathfrak{p}$-primary ideal.*

Proof. We shall apply Lemma 1 to $\mathfrak{q}$ and $\mathfrak{p}$. It is clear that $\mathfrak{q} \subseteq \mathfrak{p}$. Also, if $x \in \mathfrak{p}$ then for each i we can find an integer m_i such that $x^{m_i} \in \mathfrak{q}_i$. If, therefore, we put $m = \max(m_1, m_2, \ldots, m_n)$ we shall have $x^m \in \mathfrak{q}_i$ for $1 \leqslant i \leqslant n$, that is to say we shall have $x^m \in \mathfrak{q}$. Lastly, assume that $ab \in \mathfrak{q}$ and that $a \notin \mathfrak{p}$; then $ab \in \mathfrak{q}_i$ and $a \notin \mathfrak{p}$, hence $b \in \mathfrak{q}_i$. Since this holds for all i, b is in $\mathfrak{q}$. It follows now from Lemma 1, that $\mathfrak{q}$ is $\mathfrak{p}$-primary.

PROPOSITION 5. *If $\mathfrak{q}$ is $\mathfrak{p}$-primary and if $\mathfrak{a}$ is an ideal not contained in $\mathfrak{q}$, then $\mathfrak{q} : \mathfrak{a}$ is again $\mathfrak{p}$-primary. If, however, $\mathfrak{a} \subseteq \mathfrak{q}$ then* $\mathfrak{q} : \mathfrak{a} = (1)$.†

Proof. If $\mathfrak{a} \subseteq \mathfrak{q}$ then every element of the ring is in $\mathfrak{q} : \mathfrak{a}$, so that $\mathfrak{q} : \mathfrak{a} = R = (1)$. Suppose now that $\mathfrak{a} \nsubseteq \mathfrak{q}$ and put $\mathfrak{q}' = \mathfrak{q} : \mathfrak{a}$; we shall apply Lemma 1 to $\mathfrak{q}'$ and $\mathfrak{p}$. Since $\mathfrak{a} \nsubseteq \mathfrak{q}$ we can find $a_0 \in \mathfrak{a}$ so that $a_0 \notin \mathfrak{q}$. If now $y \in \mathfrak{q}'$ then $a_0 y \in \mathfrak{q}$ and $a_0 \notin \mathfrak{q}$, consequently $y \in \mathfrak{p}$, which proves that $\mathfrak{q}' \subseteq \mathfrak{p}$. Again, if $x \in \mathfrak{p}$ then with a suitable integer m we have $x^m \in \mathfrak{q} \subseteq \mathfrak{q}'$. Finally, assume that $\alpha\beta \in \mathfrak{q}'$ and that $\alpha \notin \mathfrak{p}$. Then for any $a \in \mathfrak{a}$ we have $a\alpha\beta \in \mathfrak{q}$ and $\alpha \notin \mathfrak{p}$, so that $a\beta \in \mathfrak{q}$, and consequently $\beta \in \mathfrak{q} : \mathfrak{a} = \mathfrak{q}'$. Lemma 1 may now be applied.

PROPOSITION 6. *Let $\mathfrak{a}$ be an ideal and let $\mathfrak{p}_1, \mathfrak{p}_2, \ldots, \mathfrak{p}_n$ be prime ideals none of which contains $\mathfrak{a}$, then there exists an element $a \in \mathfrak{a}$ such that no $\mathfrak{p}_i$ contains a.*

Proof. We use induction on the number n of prime ideals. If $n = 1$ the assertion is trivial. Let us now assume that the proposition has already been established when there are only $n - 1$ prime ideals, then for each i $(1 \leqslant i \leqslant n)$ there exists an element $a_i \in \mathfrak{a}$, which is not contained by any of $\mathfrak{p}_1, \ldots, \mathfrak{p}_{i-1}, \mathfrak{p}_{i+1}, \ldots, \mathfrak{p}_n$. If for at least one i we have $a_i \notin \mathfrak{p}_i$ there will be nothing further to prove; we can therefore confine our attention to the case in which $a_i \in \mathfrak{p}_i$ for all i. Put $a = \sum_{i=1}^{n} (a_1 a_2 \ldots a_{i-1} a_{i+1} \ldots a_n)$, then the jth term in the sum does not belong to $\mathfrak{p}_j$, but if $i \neq j$ then $a_1 \ldots a_{i-1} a_{i+1} \ldots a_n \in \mathfrak{p}_j$

† It may help the reader to fix this proposition in his mind, if he regards it as asserting that the property of being $\mathfrak{p}$-primary is not destroyed by division. To do this he must regard the ideal (1) as being, *in a conventional sense*, a primary ideal belonging to every prime ideal. The convention may seem more reasonable if he recalls that the integer 1 can be considered as a power of every prime number p; in fact $1 = p^0$.

since a_j occurs in the product. This shows that $a \notin \mathfrak{p}_j$ whatever the value of j, and as it is clear that $a \in \mathfrak{a}$—because all of $a_1, a_2, \ldots, a_n$ are in $\mathfrak{a}$—this establishes the proposition.

The union of a family of sets. Proposition 6 can be put into a more striking form, by using the concept of the union of a family of sets. Suppose that A_i, for each $i \in I$, is a set of elements of our ring (not necessarily an ideal), the range I of the suffix i being quite arbitrary. By the *union* of the A_i one then means the set of all elements x such that $x \in A_i$ for at least one $i \in I$; the union is denoted by $\bigcup_{i \in I} A_i$. In set theory union and intersection are complementary notions; in ideal theory the former concept is less important, because a union of ideals need not be an ideal while an intersection of ideals is always one. The notion of a union allows Proposition 6 to be put into the following form: *If the ideal $\mathfrak{a}$ is contained in the union of the prime ideals $\mathfrak{p}_1, \mathfrak{p}_2, \ldots, \mathfrak{p}_n$, then it is entirely contained by one of them.*

The radical of an ideal. Before we apply the results of §§ 1·2–1·5, let us notice that the construction which led from a given primary ideal to the corresponding prime ideal may be applied to a completely arbitrary ideal $\mathfrak{a}$. In fact, the set of all elements x, such that some positive power of x is in $\mathfrak{a}$, is always an ideal—the proof is exactly the same as in the primary case—which is denoted by $\operatorname{Rad} \mathfrak{a}$ and which is called the *radical* of $\mathfrak{a}$. $\operatorname{Rad} \mathfrak{a}$ will, of course, not usually be prime. It can be shown that $\operatorname{Rad}(\operatorname{Rad} \mathfrak{a}) = \operatorname{Rad} \mathfrak{a}$, and that $\operatorname{Rad}(\mathfrak{a}\mathfrak{b}) = \operatorname{Rad}(\mathfrak{a} \cap \mathfrak{b}) = \operatorname{Rad} \mathfrak{a} \cap \operatorname{Rad} \mathfrak{b}$. The proofs are quite elementary, and we shall leave their construction as an easy exercise for the reader; these relations are not needed in what follows. Closely connected with $\operatorname{Rad} \mathfrak{a}$ are the *minimal prime ideals* of $\mathfrak{a}$, which are defined as follows: *A prime ideal $\mathfrak{p}$ is called a minimal prime ideal of $\mathfrak{a}$, if it contains $\mathfrak{a}$, and if there is no prime ideal containing $\mathfrak{a}$ which is strictly contained in $\mathfrak{p}$.* The connexion between the radical of $\mathfrak{a}$ and its minimal prime ideals is explained in the following section (see Theorem 1).

1·6. Ideals with a primary decomposition. If an ideal $\mathfrak{a}$ can be expressed in the form

$$\mathfrak{a} = \mathfrak{q}_1 \cap \mathfrak{q}_2 \cap \ldots \cap \mathfrak{q}_n, \tag{1·6·1}$$

where each $\mathfrak{q}_i$ is primary, we shall say that we have a *primary decomposition* of $\mathfrak{a}$, and the individual $\mathfrak{q}_i$ will be called the *primary components* of the decomposition. There is an important type of ring—the Noetherian ring—in which every ideal has such a decomposition, and it is mainly with such rings that we shall be concerned. For the present, however, it is more convenient not to impose conditions on the ring itself, but, as occasion demands, to restrict our attention to *decomposable* ideals, as we shall call those ideals which can be written in the form (1·6·1).

THEOREM 1. *Suppose that* $\mathfrak{a}=\mathfrak{q}_1 \cap \mathfrak{q}_2 \cap \ldots \cap \mathfrak{q}_n$, *where* $\mathfrak{q}_i$ *is* $\mathfrak{p}_i$*-primary for* $1 \leqslant i \leqslant n$. *Then any prime ideal which contains* $\mathfrak{a}$ *must contain at least one of the* $\mathfrak{p}_i$; *the minimal prime ideals of* $\mathfrak{a}$ *are just those prime ideals* $\mathfrak{p}_i$ *which do not strictly contain any other* $\mathfrak{p}_j$; $\operatorname{Rad}\mathfrak{a}=\mathfrak{p}_1 \cap \mathfrak{p}_2 \cap \ldots \cap \mathfrak{p}_n$, *and, more precisely, the radical of* $\mathfrak{a}$ *is the intersection of all the minimal prime ideals of* $\mathfrak{a}$.

Proof. Let $\mathfrak{p}$ be a prime ideal containing $\mathfrak{a}$, then

$$\mathfrak{q}_1\mathfrak{q}_2\ldots\mathfrak{q}_n \subseteq \mathfrak{q}_1 \cap \mathfrak{q}_2 \cap \ldots \cap \mathfrak{q}_n = \mathfrak{a} \subseteq \mathfrak{p},$$

consequently, by Proposition 2, we can choose i so that $\mathfrak{q}_i \subseteq \mathfrak{p}$, and then, by Proposition 3, $\mathfrak{p}_i \subseteq \mathfrak{p}$. This proves the first assertion, and the second assertion follows from the first simply by applying the definition of the minimal prime ideals of $\mathfrak{a}$. Next let $x \in \operatorname{Rad}\mathfrak{a}$, then with a suitable integer m we have $x^m \in \mathfrak{a} \subseteq \mathfrak{q}_i \subseteq \mathfrak{p}_i$, so that $x \in \mathfrak{p}_i$ for $1 \leqslant i \leqslant n$. Again, if $y \in \mathfrak{p}_1 \cap \mathfrak{p}_2 \cap \ldots \cap \mathfrak{p}_n$, then for each i, $y \in \mathfrak{p}_i$, hence we can find m_i such that $y^{m_i} \in \mathfrak{q}_i$. After putting $m=\max(m_1, m_2, \ldots, m_n)$ we have $y^m \in \mathfrak{q}_1 \cap \mathfrak{q}_2 \cap \ldots \cap \mathfrak{q}_n = \mathfrak{a}$, and therefore $y \in \operatorname{Rad}\mathfrak{a}$. It has now been shown that

$$\operatorname{Rad}\mathfrak{a}=\mathfrak{p}_1 \cap \mathfrak{p}_2 \cap \ldots \cap \mathfrak{p}_n.$$

If we note that from $\mathfrak{p}_1 \cap \mathfrak{p}_2 \cap \ldots \cap \mathfrak{p}_n$ we may drop every $\mathfrak{p}_i$ which strictly contains at least one of $\mathfrak{p}_1, \mathfrak{p}_2, \ldots \mathfrak{p}_n$, this will yield the last assertion, for (by our earlier remarks) only the minimal prime ideals of $\mathfrak{a}$ will remain.

Theorem 1 shows that a decomposable ideal will have only a finite number of minimal prime ideals, and that these will be associated with every primary decomposition. We shall now consider what further properties two different decompositions of a given ideal will have in common. A little reflexion suggests that, before we make any comparisons, it will be profitable to refine—in a certain sense—the given decompositions. Suppose that $\mathfrak{a} = \mathfrak{q}_1 \cap \mathfrak{q}_2 \cap \ldots \cap \mathfrak{q}_n$, where $\mathfrak{q}_i$ is $\mathfrak{p}_i$-primary; it may happen that $\mathfrak{p}_1, \mathfrak{p}_2, \ldots, \mathfrak{p}_n$ are not all distinct. Let us suppose that

$$\mathfrak{p}_{i_1} = \mathfrak{p}_{i_2} = \ldots = \mathfrak{p}_{i_r} = \mathfrak{p},$$

then, by Proposition 4,

$$\mathfrak{q}_{i_1} \cap \mathfrak{q}_{i_2} \cap \ldots \cap \mathfrak{q}_{i_r} = \mathfrak{q} \quad \text{(say)}$$

is $\mathfrak{p}$-primary, so that we may replace all of $\mathfrak{q}_{i_1}, \mathfrak{q}_{i_2}, \ldots, \mathfrak{q}_{i_r}$ by the single primary ideal $\mathfrak{q}$. Again, if $\mathfrak{q}_i$ contains the intersection of the remaining $\mathfrak{q}_j$ it may be left out altogether. *A decomposition in which no* $\mathfrak{q}_i$ *contains the intersection of the remaining* $\mathfrak{q}_j$ *is called 'irredundant', and an irredundant decomposition, in which the prime ideals belonging to the various primary components are all different, is called a 'normal decomposition'.* Each primary decomposition can be refined into one which is normal; for if we first combine all primary ideals which belong to the same prime ideal, and then drop redundant terms one by one, we shall reach the required form in a finite number of steps. We propose now to compare two normal decompositions of the same decomposable ideal, but before doing this we shall make an observation of a rather trivial nature, in order not to encumber the proof of the next result. The ideal (1), which consists of all the elements of the ring, is a prime ideal, and there are no (1)-primary ideals other than (1) itself; for such an ideal will contain a power of 1, that is to say it will contain the unit element. It follows, therefore, that if we have an *irredundant* decomposition of a *proper* ideal, then all the prime ideals associated with the decomposition will be proper as well.

THEOREM 2. *Suppose that the ideal* $\mathfrak{a}$ *has a primary decomposition, and let* $\mathfrak{a} = \mathfrak{q}_1 \cap \mathfrak{q}_2 \cap \ldots \cap \mathfrak{q}_m = \mathfrak{q}'_1 \cap \mathfrak{q}'_2 \cap \ldots \cap \mathfrak{q}'_n$ *be two normal decompositions of* $\mathfrak{a}$, *where* $\mathfrak{q}_i$ *is* $\mathfrak{p}_i$*-primary and where* $\mathfrak{q}'_j$ *is* $\mathfrak{p}'_j$*-primary. Then* $m = n$, *and it is possible to number the components in such a way that* $\mathfrak{p}_i = \mathfrak{p}'_i$ *for* $1 \leqslant i \leqslant m = n$.

Proof. If $\mathfrak{a} = (1)$ the assertion is trivial. We may suppose therefore that $\mathfrak{a} \neq (1)$, in which case all the prime ideals

$$\mathfrak{p}_1, \ldots, \mathfrak{p}_m, \mathfrak{p}'_1, \ldots, \mathfrak{p}'_n$$

are proper. From this set of prime ideals we select one which is not strictly contained by any of the others. Without loss of generality we may suppose that the one which has been selected is $\mathfrak{p}_m$. *We assert that* $\mathfrak{p}_m$ *occurs among* $\mathfrak{p}'_1, \mathfrak{p}'_2, \ldots, \mathfrak{p}'_n$. To prove this it will be enough to show that $\mathfrak{p}_m \subseteq \mathfrak{p}'_j$ for some j, and, by Proposition 3, the latter assertion will follow if we show that $\mathfrak{q}_m \subseteq \mathfrak{p}'_j$ for some j. We shall now assume that for every j we have $\mathfrak{q}_m \nsubseteq \mathfrak{p}'_j$ and derive a contradiction. By Corollary 3 of Proposition 3 $\mathfrak{q}'_j : \mathfrak{q}_m = \mathfrak{q}'_j$, consequently, by (5) of Proposition 1,

$$\begin{aligned}\mathfrak{a} : \mathfrak{q}_m &= (\mathfrak{q}'_1 : \mathfrak{q}_m) \cap (\mathfrak{q}'_2 : \mathfrak{q}_m) \cap \ldots \cap (\mathfrak{q}'_n : \mathfrak{q}_m)\\ &= \mathfrak{q}'_1 \cap \mathfrak{q}'_2 \cap \ldots \cap \mathfrak{q}'_n = \mathfrak{a}.\end{aligned}$$

But, on one hand, if $1 \leqslant i < m$, $\mathfrak{p}_m \nsubseteq \mathfrak{p}_i$ (otherwise we should have $\mathfrak{p}_m = \mathfrak{p}_i$) and therefore $\mathfrak{q}_m \nsubseteq \mathfrak{p}_i$, so that (again by Corollary 3 of Proposition 3) $\mathfrak{q}_i : \mathfrak{q}_m = \mathfrak{q}_i$; and, on the other hand, $\mathfrak{q}_m : \mathfrak{q}_m = (1)$. These relations show that $\mathfrak{a} : \mathfrak{q}_m = \mathfrak{q}_1 \cap \mathfrak{q}_2 \cap \ldots \cap \mathfrak{q}_{m-1}$, but, since we already know that $\mathfrak{a} : \mathfrak{q}_m = \mathfrak{a}$, we have proved that

$$\mathfrak{a} = \mathfrak{q}_1 \cap \mathfrak{q}_2 \cap \ldots \cap \mathfrak{q}_{m-1}.$$

This, however, contradicts the hypothesis that the given decompositions are normal.†

Now that it has been proved that $\mathfrak{p}_m$ occurs among $\mathfrak{p}'_1, \mathfrak{p}'_2, \ldots, \mathfrak{p}'_n$ we may, without loss of generality, suppose that $\mathfrak{p}_m = \mathfrak{p}'_n$. Put $\mathfrak{q} = \mathfrak{q}_m \cap \mathfrak{q}'_n$ then, by Proposition 4, $\mathfrak{q}$ is a primary ideal belonging to $\mathfrak{p}_m = \mathfrak{p}'_n$. Also, $\mathfrak{q}_i : \mathfrak{q} = \mathfrak{q}_i$ for $1 \leqslant i < m$ and $\mathfrak{q}_m : \mathfrak{q} = (1)$—the first relation follows from the fact that since $\mathfrak{p}_m \nsubseteq \mathfrak{p}_i$, $\mathfrak{q}$ is not contained in $\mathfrak{p}_i$, while the second follows from $\mathfrak{q} \subseteq \mathfrak{q}_m$—consequently $\mathfrak{a} : \mathfrak{q} = \mathfrak{q}_1 \cap \mathfrak{q}_2 \cap \ldots \cap \mathfrak{q}_{m-1}$. An exactly similar argument shows that

$$\mathfrak{a} : \mathfrak{q} = \mathfrak{q}'_1 \cap \mathfrak{q}'_2 \cap \ldots \cap \mathfrak{q}'_{n-1},$$

hence

$$\mathfrak{q}_1 \cap \mathfrak{q}_2 \cap \ldots \cap \mathfrak{q}_{m-1} = \mathfrak{q}'_1 \cap \mathfrak{q}'_2 \cap \ldots \cap \mathfrak{q}'_{n-1},$$

and, moreover, both decompositions are normal. We have, therefore, a situation entirely similar to that with which we started, consequently we can renumber the components in such a way that we have $\mathfrak{p}_{m-1} = \mathfrak{p}'_{n-1}$ and

$$\mathfrak{q}_1 \cap \mathfrak{q}_2 \cap \ldots \cap \mathfrak{q}_{m-2} = \mathfrak{q}'_1 \cap \mathfrak{q}'_2 \cap \ldots \cap \mathfrak{q}'_{n-2}.$$

† If $m=1$ the argument must be varied slightly. In this case we obtain $\mathfrak{a} = \mathfrak{a} : \mathfrak{q}_m = (1)$ which is again a contradiction.

It is now clear that the theorem will be established if we show that $m=n$. But suppose, for example, that $m<n$, then after m steps we should obtain

$$(1)=\mathfrak{q}'_1 \cap \mathfrak{q}'_2 \cap \ldots \cap \mathfrak{q}'_{n-m} \subseteq \mathfrak{p}'_1 \cap \mathfrak{p}'_2 \cap \ldots \cap \mathfrak{p}'_{n-m},$$

which is not possible since all the $\mathfrak{p}'_j$ are proper.

Theorem 2 shows that if $\mathfrak{a}$ is decomposable then the prime ideals, which are associated with a normal decomposition of $\mathfrak{a}$, depend only on $\mathfrak{a}$ and not on the particular normal decomposition considered. These prime ideals will be called *the prime ideals belonging to* $\mathfrak{a}$. By Theorem 1, every minimal prime ideal of $\mathfrak{a}$ belongs to $\mathfrak{a}$ in this sense. The prime ideals of $\mathfrak{a}$ which are not minimal are said (for geometrical reasons) to be *embedded*.

1·7. The isolated components of an ideal. Suppose that S is a (non-empty) *multiplicatively closed* set, by which we mean that S has the property that whenever two elements belong to S their product also belongs to S. If now $\mathfrak{a}$ is an ideal we shall denote by $\mathfrak{a}_S$ the set of all elements x such that $cx \in \mathfrak{a}$ for at least one $c \in S$. $\mathfrak{a}_S$, as the reader can verify without difficulty, is an ideal; we shall call it either *the isolated component of* $\mathfrak{a}$ *determined by* S, or, more simply, *the* S*-component of* $\mathfrak{a}$. If we have a primary decomposition of $\mathfrak{a}$ it is easy to express $\mathfrak{a}_S$ in terms of the decomposition (see Proposition 7), and we shall use this fact to obtain further invariant features of normal decompositions; however, the concept of an S-component is of great use even when we are considering ideals which are not decomposable.

If we take, as we may, S to consist only of the unit element, the corresponding S-component will be $\mathfrak{a}$ itself; if we let S be the set whose only element is the zero element, $\mathfrak{a}_S$ becomes the whole ring. These are the extreme cases; all other S-components lie between these two. Before proceeding, let us note that if $\mathfrak{p}$ is a proper prime ideal, $R-\mathfrak{p}$, that is the set of all elements not in $\mathfrak{p}$, is multiplicatively closed; in fact, this is one of the most important ways in which multiplicatively closed sets arise.

PROPOSITION 7. *Suppose that* $\mathfrak{a}=\mathfrak{q}_1 \cap \mathfrak{q}_2 \cap \ldots \cap \mathfrak{q}_n$, *where* $\mathfrak{q}_i$ *is* $\mathfrak{p}_i$*-primary. Let* $\mathfrak{p}_i$ *meet* S *for* $m+1 \leqslant i \leqslant n$ *but not for* $1 \leqslant i \leqslant m$, *then* $\mathfrak{a}_S=\mathfrak{q}_1 \cap \mathfrak{q}_2 \cap \ldots \cap \mathfrak{q}_m$.

Proof. Let $x \in \mathfrak{a}_S$ then $cx \in \mathfrak{a} = \mathfrak{q}_1 \cap \mathfrak{q}_2 \cap \ldots \cap \mathfrak{q}_n$, where c is a suitable element of S. Consequently, if $1 \leqslant i \leqslant m$ we have $cx \in \mathfrak{q}_i$ and $c \notin \mathfrak{p}_i$, which shows that $x \in \mathfrak{q}_i$ for $1 \leqslant i \leqslant m$. Now let

$$y \in \mathfrak{q}_1 \cap \mathfrak{q}_2 \cap \ldots \cap \mathfrak{q}_m.$$

For each $j > m$ we can choose $c_j \in \mathfrak{p}_j \cap S$, and then, if N is sufficiently large, $(c_{m+1}c_{m+2}\ldots c_n)^N \in \mathfrak{q}_{m+1} \cap \mathfrak{q}_{m+2} \cap \ldots \cap \mathfrak{q}_n$; hence $y(c_{m+1}c_{m+2}\ldots c_n)^N \in \mathfrak{q}_1 \cap \ldots \cap \mathfrak{q}_n = \mathfrak{a}$. But, since S is multiplicatively closed, $(c_{m+1}c_{m+2}\ldots c_n)^N \in S$ and therefore $y \in \mathfrak{a}_S$. This completes the proof.

COROLLARY. *A decomposable ideal has at most a finite number of isolated components.*

Suppose that the ideal $\mathfrak{a}$ is decomposable, and let $\mathfrak{p}_1, \mathfrak{p}_2, \ldots, \mathfrak{p}_n$ be the prime ideals which belong to $\mathfrak{a}$. A subset $\mathfrak{p}_{i_1}, \mathfrak{p}_{i_2}, \ldots, \mathfrak{p}_{i_r}$ of these prime ideals is called an *isolated set of prime ideals belonging to* $\mathfrak{a}$, if every $\mathfrak{p}_j$, which is contained by at least one of the subset $\mathfrak{p}_{i_1}, \mathfrak{p}_{i_2}, \ldots, \mathfrak{p}_{i_r}$, is necessarily a member of the subset itself. For examples, each minimal prime ideal of $\mathfrak{a}$ forms on its own an isolated set of prime ideals belonging to $\mathfrak{a}$; and, again, the $\mathfrak{p}_j$ which do not meet a given set of elements will also form an isolated set of prime ideals of $\mathfrak{a}$. We come now to the second uniqueness theorem for normal decompositions.

THEOREM 3. *Suppose that* $\mathfrak{a}$ *has a primary decomposition, and let* $\mathfrak{a} = \mathfrak{q}_1 \cap \mathfrak{q}_2 \cap \ldots \cap \mathfrak{q}_n$ *be a normal decomposition of* $\mathfrak{a}$, *where* $\mathfrak{q}_i$ *is* $\mathfrak{p}_i$*-primary. If now* $\mathfrak{p}_{i_1}, \mathfrak{p}_{i_2}, \ldots, \mathfrak{p}_{i_r}$ *is an isolated set of prime ideals belonging to* $\mathfrak{a}$, *then* $\mathfrak{q}_{i_1} \cap \mathfrak{q}_{i_2} \cap \ldots \cap \mathfrak{q}_{i_r}$ *depends only on* $\mathfrak{p}_{i_1}, \mathfrak{p}_{i_2}, \ldots, \mathfrak{p}_{i_r}$, *and not on the particular normal decomposition considered.*

Proof. Let S consist of all elements not contained by any of $\mathfrak{p}_{i_1}, \mathfrak{p}_{i_2}, \ldots, \mathfrak{p}_{i_r}$, then S is multiplicatively closed. We shall prove that $\mathfrak{a}_S = \mathfrak{q}_{i_1} \cap \mathfrak{q}_{i_2} \cap \ldots \cap \mathfrak{q}_{i_r}$, from which the theorem itself will follow, since $\mathfrak{a}_S$ is determined solely by $\mathfrak{a}$ and the prime ideals $\mathfrak{p}_{i_1}, \mathfrak{p}_{i_2}, \ldots, \mathfrak{p}_{i_r}$. By Proposition 7, it is only necessary to show that every $\mathfrak{p}_j$, which is not one of $\mathfrak{p}_{i_1}, \mathfrak{p}_{i_2}, \ldots, \mathfrak{p}_{i_r}$, meets S. But, since $\mathfrak{p}_{i_1}, \mathfrak{p}_{i_2}, \ldots, \mathfrak{p}_{i_r}$ is an isolated set, for such a $\mathfrak{p}_j$ we shall have $\mathfrak{p}_j \nsubseteq \mathfrak{p}_{i_1}$, $\mathfrak{p}_j \nsubseteq \mathfrak{p}_{i_2}, \ldots, \mathfrak{p}_j \nsubseteq \mathfrak{p}_{i_r}$, consequently, by Proposition 6, $\mathfrak{p}_j$ contains an element not in any of $\mathfrak{p}_{i_1}, \mathfrak{p}_{i_2}, \ldots, \mathfrak{p}_{i_r}$. In other words, $\mathfrak{p}_j$ meets S.

COROLLARY. *Let $\mathfrak{p}$ be a minimal prime ideal of $\mathfrak{a}$, then the primary component corresponding to $\mathfrak{p}$ is the same for all normal decompositions of $\mathfrak{a}$.*

We shall show later, by means of an example, that the corresponding assertion for embedded prime ideals is false.

1·8. Noetherian rings. The Noetherian rings, which we are about to introduce, are important for at least three reasons. First, in such a ring every ideal is decomposable. In order to apply the primary decomposition to the deeper problems of ideal theory, we shall use a technique, by means of which we transform the ring with which we start into one in which the problem under discussion becomes simplified. The second reason why Noetherian rings are important is that the Noetherian property is not destroyed by these transformations. On a less abstract level, ideal theory is used as the main tool in attacking the problem of providing algebraic geometry with adequate algebraic foundations. A third reason why Noetherian rings are important, is that the rings, which arise naturally in connexion with algebraic geometry, are almost always of this type.

We recall that an ideal $\mathfrak{a}$ is finitely generated (see §1·3), if we can find a finite set $a_1, a_2, \ldots, a_n$ of elements, such that

$$\mathfrak{a} = Ra_1 + Ra_2 + \ldots + Ra_n.$$

Ra_1 is just another way of writing the principal ideal (a_1), which is used when we wish to emphasize that (a_1) consists of all elements of the form ra_1, where r is an arbitrary element of R.

DEFINITION. *A ring R is called 'Noetherian' if every ideal of R is finitely generated.*

In order to put the Noetherian condition into alternative forms, we make two further definitions.

DEFINITION. *The 'ascending chain' condition is said to hold in R, if, whenever we have an infinite increasing sequence*

$$\mathfrak{a}_1 \subseteq \mathfrak{a}_2 \subseteq \mathfrak{a}_3 \subseteq \ldots$$

of ideals, there exists an integer m such that $\mathfrak{a}_n = \mathfrak{a}_m$ for all $n \geqslant m$.

DEFINITION. *The 'maximal condition' is said to hold in R, if given any non-empty set $\mathfrak{S}$ of ideals, there is always an ideal $\mathfrak{a}$, in the set $\mathfrak{S}$, and such that if $\mathfrak{b}$ belongs to $\mathfrak{S}$ and $\mathfrak{b} \supseteq \mathfrak{a}$, then $\mathfrak{b} = \mathfrak{a}$.*

A more casual way of wording the definition, is to say that every non-empty set $\mathfrak{S}$ of ideals, contains one, $\mathfrak{a}$ (say), which is maximal. This, however, does not mean that the ideal $\mathfrak{a}$, which is maximal, contains every ideal of the set $\mathfrak{S}$; it only means that $\mathfrak{a}$ is not contained by any other ideal of the set.

THEOREM 4. *The following three statements are equivalent:*

(1) *The ascending chain condition holds in R;*

(2) *The maximal condition, for ideals, holds in R;*

(3) *Every ideal in R is finitely generated;*

and each of these is equivalent to: 'R is Noetherian'.

Proof. Assume that the chain condition holds, and let $\mathfrak{S}$ be a non-empty set of ideals. We shall suppose that no member of $\mathfrak{S}$ is maximal, and hence derive a contradiction. This will prove that (1) implies (2). Since $\mathfrak{S}$ is not empty, it contains at least one ideal; let $\mathfrak{a}_1 \in \mathfrak{S}$. By hypothesis, $\mathfrak{a}_1$ cannot be maximal; we can therefore find $\mathfrak{a}_2 \in \mathfrak{S}$ such that $\mathfrak{a}_2 \supset \mathfrak{a}_1$, where the inclusion is strict. Again, since $\mathfrak{a}_2$ is not maximal, there exists $\mathfrak{a}_3 \in \mathfrak{S}$ such that $\mathfrak{a}_3 \supset \mathfrak{a}_2$—and so on. This gives the required contradiction, because the sequence $\mathfrak{a}_1, \mathfrak{a}_2, \ldots$ will violate the chain condition.

Now assume that the maximal condition holds, and let $\mathfrak{a}$ be a given ideal. Denote by $\mathfrak{S}$ the set of all finitely generated ideals contained in $\mathfrak{a}$, then $\mathfrak{S}$ is not empty because it contains (0). If $\mathfrak{a}^* = Ra_1 + Ra_2 + \ldots + Ra_n$ is an ideal of the set $\mathfrak{S}$ which is maximal, then $\mathfrak{a}^* \subseteq \mathfrak{a}$. We shall prove that $\mathfrak{a}^* = \mathfrak{a}$, from which it will follow that $\mathfrak{a}$ is finitely generated, and, hence, that (2) implies (3). Now if $\mathfrak{a}^* \neq \mathfrak{a}$, we can find $b \in \mathfrak{a}$ such that $b \notin \mathfrak{a}^*$, and then the ideal

$$Ra_1 + Ra_2 + \ldots + Ra_n + Rb$$

will belong to $\mathfrak{S}$, and it will strictly contain $\mathfrak{a}^*$. This, however, is impossible by the choice of $\mathfrak{a}^*$.

We shall complete the proof of the theorem by showing that (3) implies (1). For this, assume that every ideal is finitely generated, and let $\mathfrak{a}_1 \subseteq \mathfrak{a}_2 \subseteq \ldots$ be an increasing sequence of ideals.

If we denote by $\mathfrak{a}$ the union† of all the $\mathfrak{a}_i$, then $\mathfrak{a}$ is an ideal; for if $x_1, x_2 \in \mathfrak{a}$ and $r \in R$, then we can find an integer l such that x_1 and x_2 are *both* in $\mathfrak{a}_l$, hence $x_1 + x_2$, $x_1 - x_2$, and rx_1 are all in $\mathfrak{a}_l$, so that *a fortiori* they are all in $\mathfrak{a}$. Since $\mathfrak{a}$ is an ideal, it has, by hypothesis, a finite base. Let $\mathfrak{a} = (a_1, a_2, \ldots, a_n)$ and for each i choose m_i so that $a_i \in \mathfrak{a}_{m_i}$, then all the a_i are in $\mathfrak{a}_m$, where

$$m = \max(m_1, m_2, \ldots, m_n).$$

If now $n > m$ we have

$$\mathfrak{a} = (a_1, a_2, \ldots, a_n) \subseteq \mathfrak{a}_m \subseteq \mathfrak{a}_n \subseteq \mathfrak{a};$$

thus $\mathfrak{a}_n = \mathfrak{a}_m$ provided only that $n > m$.

We come now to one of the fundamental properties of Noetherian rings.

THEOREM 5. *Every ideal of a Noetherian ring has a primary decomposition.*

To prove this result, we shall introduce an auxiliary concept. We shall say that an ideal $\mathfrak{a}$ is *irreducible*, if whenever $\mathfrak{a} = \mathfrak{b} \cap \mathfrak{c}$ ($\mathfrak{b}$ and $\mathfrak{c}$ are ideals) then either $\mathfrak{a} = \mathfrak{b}$ or $\mathfrak{a} = \mathfrak{c}$. Theorem 5 is then an immediate consequence of the following two lemmas:

LEMMA 2. *If R is Noetherian, then every ideal can be represented as the intersection of a finite number of irreducible ideals.*

LEMMA 3. *If R is Noetherian, then every irreducible ideal is primary.*

Proof of Lemma 2. Let $\mathfrak{S}$ be the set of all ideals which are not finite intersections of irreducible ideals. We have to show that $\mathfrak{S}$ is empty. Assuming the contrary, we can (by Theorem 4) find an ideal $\mathfrak{a} \in \mathfrak{S}$, which is maximal for the set $\mathfrak{S}$. Since $\mathfrak{a} \in \mathfrak{S}$, it is not a finite intersection of irreducible ideals, so that, in particular, $\mathfrak{a}$ is not irreducible. Thus $\mathfrak{a} = \mathfrak{b} \cap \mathfrak{c}$, where $\mathfrak{b}$ and $\mathfrak{c}$ are ideals, which strictly contain $\mathfrak{a}$. By the maximal property of $\mathfrak{a}$, $\mathfrak{b} \notin \mathfrak{S}$ and $\mathfrak{c} \notin \mathfrak{S}$, consequently $\mathfrak{b}$ and also $\mathfrak{c}$ are finite intersections of irreducible ideals. It follows that $\mathfrak{a} = \mathfrak{b} \cap \mathfrak{c}$ is also a finite intersection of irreducible ideals; but this is impossible because $\mathfrak{a} \in \mathfrak{S}$.

Proof of Lemma 3. We shall suppose that $\mathfrak{a}$ is a non-primary ideal in our Noetherian ring, and we shall deduce that $\mathfrak{a}$ must be

† See §1·5.

reducible. Since $\mathfrak{a}$ is not primary, there exist elements b, c such that $bc \in \mathfrak{a}$, $c \notin \mathfrak{a}$, and no power of b is in $\mathfrak{a}$. From $bc \in \mathfrak{a}$ and $c \notin \mathfrak{a}$, it follows that $\mathfrak{a} \subset \mathfrak{a} : (b)$, where the inclusion is strict. Using (4) and (6) of Proposition 1 we obtain

$$\mathfrak{a} : (b^r) \subseteq [\mathfrak{a} : (b^r)] : (b) = \mathfrak{a} : (b^{r+1}),$$

hence

$$\mathfrak{a} \subset \mathfrak{a} : (b) \subseteq \mathfrak{a} : (b^2) \subseteq \mathfrak{a} : (b^3) \subseteq \ldots$$

The chain condition now shows that there exists an integer m such that $\mathfrak{a} : (b^n) = \mathfrak{a} : (b^m)$, provided $n > m$. We shall prove that

$$\mathfrak{a} = [\mathfrak{a} : (b^m)] \cap [\mathfrak{a} + (b^m)], \tag{A}$$

and this will establish the lemma, for, by construction, both $\mathfrak{a} : (b^m)$ and $\mathfrak{a} + (b^m)$ strictly contain $\mathfrak{a}$. Let

$$x \in [\mathfrak{a} : (b^m)] \cap [\mathfrak{a} + (b^m)],$$

then (A) will be proved if we show that $x \in \mathfrak{a}$. Now, since $x \in \mathfrak{a} + (b^m)$, we have $x = a + rb^m$, where $a \in \mathfrak{a}$ and $r \in R$. We also have $x \in \mathfrak{a} : (b^m)$; accordingly $xb^m = ab^m + rb^{2m}$ belongs to $\mathfrak{a}$, which shows that $rb^{2m} \in \mathfrak{a}$, and therefore that $r \in \mathfrak{a} : (b^{2m})$. By the choice of m, $\mathfrak{a} : (b^{2m}) = \mathfrak{a} : (b^m)$, hence $r \in \mathfrak{a} : (b^m)$ and $rb^m \in \mathfrak{a}$. Thus $x = a + rb^m \in \mathfrak{a}$, which is what we had to prove.

1·9. Some additional properties of Noetherian rings. We shall collect here certain results of an elementary nature, which will be needed in the more advanced theory of Noetherian rings that is to be developed in Chapter III.

PROPOSITION 8. *In a Noetherian ring every ideal contains a power of its radical.*

Proof. Let $\mathfrak{a}$ be an ideal in a Noetherian ring and let $\mathfrak{b} = \text{Rad}\,\mathfrak{a}$; then $\mathfrak{b}$ is finitely generated, say $\mathfrak{b} = Rb_1 + Rb_2 + \ldots + Rb_n$. Since $b_i \in \text{Rad}\,\mathfrak{a}$, we can find an integer m_i such that $b_i^{m_i} \in \mathfrak{a}$. Put $m = m_1 + m_2 + \ldots + m_n$. We shall show that $\mathfrak{b}^m \subseteq \mathfrak{a}$. Now $\mathfrak{b}^m$ is generated by the elements $b_1^{\mu_1} b_2^{\mu_2} \ldots b_n^{\mu_n}$, where the μ_i are non-negative integers such that $\mu_1 + \mu_2 + \ldots + \mu_n = m_1 + m_2 + \ldots + m_n$; but if $\mu_1 + \mu_2 + \ldots + \mu_n = m_1 + m_2 + \ldots + m_n$, then for at least one value of i, $\mu_i \geqslant m_i$, consequently $b_1^{\mu_1} b_2^{\mu_2} \ldots b_n^{\mu_n} \in \mathfrak{a}$. Since all the generators of $\mathfrak{b}^m$ are in $\mathfrak{a}$, $\mathfrak{b}^m \subseteq \mathfrak{a}$.

COROLLARY. *If R is Noetherian and if $\mathfrak{q}$ is $\mathfrak{p}$-primary, then $\mathfrak{p}^\rho \subseteq \mathfrak{q}$ for some positive integer ρ.*

The corollary shows that if $\mathfrak{p}$ is a prime ideal in a Noetherian ring, a necessary condition that a given ideal should be $\mathfrak{p}$-primary is that the ideal should contain a power of $\mathfrak{p}$. There is one important situation in which the condition is sufficient as well as necessary. Let us make the following

DEFINITION. *A proper prime ideal $\mathfrak{p}$ is said to be a 'maximal prime ideal' of the ring R, if there is no other proper prime ideal containing $\mathfrak{p}$.*

PROPOSITION 9. *Let R be a Noetherian ring and let $\mathfrak{p}$ be a maximal prime ideal of R. Then a proper ideal $\mathfrak{a}$ is $\mathfrak{p}$-primary if, and only if, it contains a power of $\mathfrak{p}$.*

Proof. By Proposition 8, it is enough to prove that if $\mathfrak{p}^r \subseteq \mathfrak{a}$ then $\mathfrak{a}$ is $\mathfrak{p}$-primary. Let $\mathfrak{p}'$ be a prime ideal belonging to $\mathfrak{a}$, then, since $\mathfrak{a}$ is proper, $\mathfrak{p}'$ is also proper.† We have $\mathfrak{p}^r \subseteq \mathfrak{a} \subseteq \mathfrak{p}'$, so that $\mathfrak{p} \subseteq \mathfrak{p}'$, and hence, since $\mathfrak{p}$ is maximal, $\mathfrak{p} = \mathfrak{p}'$. Thus $\mathfrak{p}$ is the only prime ideal belonging to $\mathfrak{a}$, or, in other words, $\mathfrak{a}$ is $\mathfrak{p}$-primary.

THEOREM 6. *Suppose that R is Noetherian, that $\mathfrak{a}$ is a proper ideal, and that $\mathfrak{b}$ is an arbitrary ideal. Then $\mathfrak{a} : \mathfrak{b} = \mathfrak{a}$ if, and only if, $\mathfrak{b}$ is not contained by any prime ideal belonging to $\mathfrak{a}$.*

Proof. Let $\mathfrak{a} = \mathfrak{q}_1 \cap \mathfrak{q}_2 \cap \ldots \cap \mathfrak{q}_n$ be a normal decomposition of $\mathfrak{a}$, where $\mathfrak{q}_i$ is $\mathfrak{p}_i$-primary. First assume that no $\mathfrak{p}_i$ contains $\mathfrak{b}$, then, by Corollary 3 of Proposition 3, $\mathfrak{q}_i : \mathfrak{b} = \mathfrak{q}_i$ for every i, hence $\mathfrak{a} : \mathfrak{b} = \mathfrak{a}$ as required.

Next, assuming that $\mathfrak{a} : \mathfrak{b} = \mathfrak{a}$, and that $\mathfrak{b}$ is contained by some $\mathfrak{p}_i$, say $\mathfrak{b} \subseteq \mathfrak{p}_1$, we shall obtain a contradiction. Dividing the equation $\mathfrak{a} : \mathfrak{b} = \mathfrak{a}$ by $\mathfrak{b}$ and using (6) of Proposition 1, we find $\mathfrak{a} = \mathfrak{a} : \mathfrak{b} = \mathfrak{a} : \mathfrak{b}^2$; and generally we have $\mathfrak{a} = \mathfrak{a} : \mathfrak{b}^r$ for all r. But $\mathfrak{b}^r \subseteq \mathfrak{p}_1^r$, consequently, by the corollary of Proposition 8, we can choose r so that $\mathfrak{b}^r \subseteq \mathfrak{q}_1$. We then have

$$\mathfrak{a} = (1) \cap (\mathfrak{q}_2 : \mathfrak{b}^r) \cap \ldots \cap (\mathfrak{q}_n : \mathfrak{b}^r),$$

hence, since $\mathfrak{a}$ is proper, $n \geqslant 2$ and

$$\mathfrak{a} = (\mathfrak{q}_2 : \mathfrak{b}^r) \cap \ldots \cap (\mathfrak{q}_n : \mathfrak{b}^r).$$

† See the remark just preceding Theorem 2 in § 1.6.

Now each of the terms in

$$(\mathfrak{q}_2 : \mathfrak{b}^r) \cap \ldots \cap (\mathfrak{q}_n : \mathfrak{b}^r)$$

is either primary or else it is the whole ring (see Proposition 5), and not every term can be the whole ring. If we keep only those terms which are proper, we obtain for $\mathfrak{a}$ a primary decomposition with which the prime ideal $\mathfrak{p}_1$ is not associated. This primary decomposition can be refined into a normal decomposition of $\mathfrak{a}$, with which $\mathfrak{p}_1$ is not associated; but this, by Theorem 2, is impossible.

1·10. Some different kinds of rings. It is clear that the positive and negative integers form a ring if we allow *addition* and *multiplication* to have their usual meanings. We shall now show that this ring is not only Noetherian, but also that it is *a principal ideal ring.*

THEOREM 7. *If I is the ring of positive and negative integers, then every ideal of I is a principal ideal.*

Proof. Let $\mathfrak{a}$ be an ideal of I. Since we wish to prove that $\mathfrak{a}$ is principal we may suppose that $\mathfrak{a} \neq (0)$. Let $m \neq 0$ belong to $\mathfrak{a}$, then $-m$ also belongs to $\mathfrak{a}$, from which we see that $\mathfrak{a}$ contains at least one strictly positive integer. Let d be the smallest positive integer contained in $\mathfrak{a}$. We shall prove that $\mathfrak{a} = (d)$, and for this, since it is clear that $(d) \subseteq \mathfrak{a}$, it will be enough to prove that $\mathfrak{a} \subseteq (d)$. Suppose that $a \in \mathfrak{a}$, then we have $a = dm + r$, where m and r are integers and $0 \leqslant r < d$. Since $a \in \mathfrak{a}$ and $d \in \mathfrak{a}$, we see that $r \in \mathfrak{a}$, hence, as $0 \leqslant r < d$, it follows, from the definition of d, that $r = 0$. Thus $a = dm \in (d)$, as required.

If $\mathfrak{a}$ is an ideal of I, then we can find an integer m which generates $\mathfrak{a}$. Thus $\mathfrak{a} = (m) = (-m)$, and it is clear that there are no other generators. By taking the non-negative generator in each case, we set up a 1-1 correspondence between the ideals and the non-negative integers, and this correspondence enables us to interpret our definitions and theorems, in terms of arithmetical properties of the integers. This interpretation will be found in the first of the examples at the end of the chapter.

DEFINITION. *An element u of a ring R is called a 'unit', if there exists an element x such that $ux = 1$.*

We note that x, if it exists, is unique; for from $ux=uy=1$ we obtain $y=y1=yux=1x=x$. The element x is usually denoted by u^{-1}; it is, of course, itself a unit.

DEFINITION. *A ring in which every non-zero element is a unit is called a 'field'.*

The most familiar fields are the rational numbers, the real numbers, and the complex numbers. The proposition which follows gives a condition for a ring to be a field, in terms of ideals.

PROPOSITION 10. *A necessary and sufficient condition that a ring R should be a field is that* (0) *and* (1) *should be the only ideals.*

Proof. Suppose, first, that R is a field and let $\mathfrak{a}$ be an ideal different from (0). Choose $a \in \mathfrak{a}$ so that $a \neq 0$, then $1=a^{-1}a \in \mathfrak{a}$, hence $\mathfrak{a}=(1)$. Now suppose that R is a ring whose only ideals are (0) and (1), and let x be an element which is not zero. Since $x \in (x)$, $(x) \neq (0)$, consequently $(x)=(1)$ and therefore $1 \in (x)$. But this means that there exists an element y such that $xy=1$.

Proposition 10 enables us to make the elementary but important observation that a field is a very special kind of Noetherian ring.

DEFINITION. *An element c is called a 'divisor of zero' if there exists an element $x \neq 0$ such that $cx=0$. A ring in which 0 is the only divisor of zero is called an 'integral domain'.*

A unit u can never be a divisor of zero (for from $ux=0$ it follows that $0=u^{-1}0=u^{-1}ux=x$), which shows that a field is necessarily an integral domain. Again, in an integral domain, $ab=ac$ and $a \neq 0$ together imply that $b=c$, which means that cancellation is permissible. It is sometimes said that an integral domain is a ring which has no divisors of zero. Although this is not quite correct (according to the above definition), the expression is useful and does not lead to confusion. The most familiar integral domain is the ring of rational integers.

Polynomial rings. If R is a given ring, we can consider formal expressions of the kind $a_0+a_1x+a_2x^2+\ldots+a_nx^n$, where the a_i are elements of R, and where x is a *new symbol*, which is referred to sometimes as an *indeterminate* and sometimes as a *variable*. An expression such as $a_0+a_1x+\ldots+a_nx^n$ is called a *polynomial in the variable* x. By the *coefficient of* x^i in this polynomial we mean a_i if $i \leqslant n$, and zero if $i > n$. Two polynomials

$$a_0 + a_1x + \ldots + a_nx^n \quad \text{and} \quad b_0 + b_1x + \ldots + b_px^p$$

are considered to be equal,† if, and only if, the coefficient of x^i is the same in both polynomials for all values of i. Addition and multiplication of polynomials are now defined in the natural way, and this turns the set of all polynomials in x (with coefficients in R) into a ring, which it is customary to denote by $R[x]$. The zero element of this new ring is the so-called *null polynomial*, which has all its coefficients equal to zero. The *constant* polynomials, by which we mean those polynomials which have the coefficient of x^i equal to zero for every $i \geqslant 1$, form by themselves a ring. To each element $a \in R$ there corresponds a unique constant polynomial $a + 0x + 0x^2 + \ldots$, and this correspondence shows that the ring of constant polynomials is just a copy of the ring R. We therefore identify R with the ring of constant polynomials, and say that $R[x]$ contains R. The unit element of $R[x]$ is the constant polynomial 1, or, according to our identifications, it is simply the unit element of R. By the *leading coefficient* of $f(x) = a_0 + a_1x + \ldots + a_nx^n$ we mean the last non-zero coefficient, and by the *degree* of $f(x)$ we mean the largest value of i for which the coefficient of x^i is not zero.‡ The degree of $f(x)$ will be denoted by $\partial^0 f(x)$, or by $\partial^0 f$.

So far, we have spoken only of polynomials in one variable. We may also consider polynomials in n variables $x_1, x_2, \ldots, x_n$ with coefficients in R, in which case we obtain the ring that is denoted by $R[x_1, x_2, \ldots, x_n]$. This matter of the formation of polynomials has not been treated in great detail, because the main features of the construction are so familiar that the reader will be able to proceed without further explanations.

THEOREM 8 (*Hilbert's basis theorem*). *If R is a Noetherian ring, then the polynomial ring $R[x]$ is also Noetherian.*

Proof. We shall suppose that $\mathfrak{A}$ is an ideal of $R[x]$, and we shall show that $\mathfrak{A}$ is finitely generated. The elements of $\mathfrak{A}$ are poly-

† Note that equality of polynomials is defined without the polynomials being treated, in any way, as functions.

‡ We can either leave the degree of the null polynomial undefined, or else we can regard its degree as being 'minus infinity'.

nomials. We form a set $\mathfrak{a}$, of elements of R, by taking the leading coefficients of all the polynomials in $\mathfrak{A}$ together with the zero element. Now $\mathfrak{a}$ is an ideal of R. For suppose that $\alpha_1, \alpha_2 \in \mathfrak{a}$, then there exist polynomials $\alpha_1 x^m + \dots$ and $\alpha_2 x^n + \dots$ which are in $\mathfrak{A}$, where in each case $+\dots$ stands for: *plus terms of lower degree.* Put $p = m + n$, multiply the first polynomial by x^n and the second by x^m. We thus obtain two polynomials $\alpha_1 x^p + \dots$ and $\alpha_2 x^p + \dots$ both of which are in $\mathfrak{A}$, consequently $(\alpha_1 x^p + \dots) \pm (\alpha_2 x^p + \dots) \in \mathfrak{A}$, hence $\alpha_1 \pm \alpha_2 \in \mathfrak{a}$. Again, if $r \in R$, then $r(\alpha_1 x^p + \dots) \in \mathfrak{A}$ and therefore $r\alpha_1 \in \mathfrak{a}$. It has now been proved that $\mathfrak{a}$ is an ideal. Since R is Noetherian, $\mathfrak{a}$ is finitely generated, say $\mathfrak{a} = (\alpha_1, \alpha_2, \dots, \alpha_h)$, so that there exist polynomials $f_1 = f_1(x), f_2 = f_2(x), \dots, f_h = f_h(x)$ such that $f_i \in \mathfrak{A}$, and such that $f_i(x)$ has leading coefficient α_i. By multiplying each of the $f_i(x)$ by an appropriate power of x, we can arrange that they all have the same degree, say the degree N, thus

$$f_i \in \mathfrak{A};\ f_i(x) = \alpha_i x^N + \dots \quad (1 \leqslant i \leqslant h).$$

Next, consider all the polynomials in $\mathfrak{A}$ whose degrees do not exceed $N-1$. The coefficients of x^{N-1} in these polynomials form an ideal $\mathfrak{b}$ of R, say $\mathfrak{b} = (\beta_1, \beta_2, \dots, \beta_k)$, and we can choose polynomials $g_1(x), g_2(x), \dots, g_k(x)$ such that

$$g_i \in \mathfrak{A};\ g_i(x) = \beta_i x^{N-1} + \dots \quad (1 \leqslant i \leqslant k).$$

In the same way, if we consider the polynomials in $\mathfrak{A}$ whose degrees do not exceed $N-2$, then the coefficients of x^{N-2} in these polynomials will form an ideal $\mathfrak{c} = (\gamma_1, \gamma_2, \dots, \gamma_l)$. Also there will exist polynomials $h_1(x), h_2(x), \dots, h_l(x)$ such that

$$h_i \in \mathfrak{A};\ h_i(x) = \gamma_i x^{N-2} + \dots \quad (1 \leqslant i \leqslant l).$$

By this method we eventually obtain a certain *finite* set $f_1, \dots, f_h, g_1, \dots, g_k, h_1, \dots, h_l, \dots$ of polynomials. These polynomials are all in $\mathfrak{A}$; we shall now show that, in fact, they generate $\mathfrak{A}$. Suppose that $\phi(x) = \alpha x^P + \dots$ belongs to $\mathfrak{A}$, then $\alpha \in \mathfrak{a}$, say

$$\alpha = \omega_1 \alpha_1 + \omega_2 \alpha_2 + \dots + \omega_h \alpha_h,$$

where $\omega_i \in R$. If $P \geqslant N$, then

$$\phi - \omega_1 x^{P-N} f_1 - \omega_2 x^{P-N} f_2 - \dots - \omega_h x^{P-N} f_h$$

is again in $\mathfrak{A}$, but it has a smaller degree than ϕ. If the degree of the new polynomial is still not less than N, we can reduce it further by the same device. In this way we see that there exist polynomials $A_1(x), A_2(x), \ldots, A_h(x)$ such that

$$\phi(x) = A_1(x)f_1(x) + A_2(x)f_2(x) + \ldots + A_h(x)f_h(x) + \psi(x),$$

where $\psi(x) \in \mathfrak{A}$, and $\partial^0\psi \leqslant N-1$. We shall complete the proof by showing that

$$\psi(x) = \mu_1 g_1(x) + \ldots + \mu_k g_k(x) + \nu_1 h_1(x) + \ldots + \nu_l h_l(x) + \ldots,$$

where $\mu_1, \ldots, \mu_k, \nu_1, \ldots, \nu_l, \ldots$ are all in R. To see this, first choose $\mu_1, \mu_2, \ldots, \mu_k$ so that $\psi(x)$ and $\mu_1 g_1(x) + \mu_2 g_2(x) + \ldots + \mu_k g_k(x)$ have the same coefficient of x^{N-1}. This is possible, since $\psi(x) \in \mathfrak{A}$ and $\partial^0\psi \leqslant N-1$, by the definition of the $g_i(x)$. Next choose $\nu_1, \nu_2, \ldots, \nu_l$ so that the coefficient of x^{N-2} is the same in

$$\psi(x) - \mu_1 g_1(x) - \ldots - \mu_k g_k(x)$$

as it is in

$$\nu_1 h_1(x) + \nu_2 h_2(x) + \ldots + \nu_l h_l(x).$$

The remaining steps are obvious.

COROLLARY 1. *If R is a Noetherian ring, then the polynomial ring $R[x_1, x_2, \ldots, x_n]$ is also Noetherian.*

Proof. Put $R_0 = R$, and $R_i = R[x_1, x_2, \ldots, x_i]$ for $1 \leqslant i \leqslant n$. Since every polynomial in $x_1, x_2, \ldots, x_{i+1}$ can be regarded, in precisely one way, as a polynomial in x_{i+1} whose coefficients are polynomials in $x_1, x_2, \ldots, x_i$, the ring R_{i+1} is none other than the polynomial ring $R_i[x_{i+1}]$. Accordingly, by the theorem just proved, R_{i+1} is Noetherian whenever R_i is Noetherian. But, by hypothesis, R_0 is Noetherian, consequently all the R_i are Noetherian, and, in particular, this is true of $R_n = R[x_1, x_2, \ldots, x_n]$.

COROLLARY 2. *If F is a field, then the polynomial ring*

$$F[x_1, x_2, \ldots, x_n]$$

is Noetherian.

EXAMPLES

Example 1. In the ring of rational integers every ideal is a principal ideal. The sum $(m)+(n)$ of the two ideals (m) and (n) consists of all the integers of the form $rm + sn$, consequently $(m)+(n) = (d)$, where d is the greatest common divisor of m and n. The product $(m)(n)$ is just (mn); the intersection $(m) \cap (n)$ consists of all integers divisible both by m and by n, i.e. $(m) \cap (n) = (l)$, where l is the least common multiple of m and n.

To interpret $(m):(n)$ we apply (8) of Proposition 1, which shows that $(m):(n)=(m):(d)=(m/d)$, where, as before, d is the greatest common divisor of m and n. We see now that $(m):(n)=(m)$ if, and only if, $d=1$; that is to say $(m):(n)=(m)$ if, and only if, m and n are relatively prime.

If p is a prime number then (p) is a prime ideal, for 'p divides ab' implies either that p divides a, or else that p divides b. We obtain all the prime ideals by taking the ideals (p), which are generated by prime numbers, together with the ideals (0) and (1). The prime ideals of the form (p) are all of them maximal prime ideals; and the (p)-primary ideals are just the ideals (p^r), where r is positive. Now an arbitrary positive integer m can be written as a product of powers of distinct prime numbers, say $m=p_1^{h_1}p_2^{h_2}\dots p_r^{h_r}$, and then an integer x will be divisible by m, if, and only if, it is divisible by each factor $p_i^{h_i}$. Thus $(m)=(p_1^{h_1})\cap(p_2^{h_2})\cap\dots\cap(p_r^{h_r})$, and this is the normal decomposition of (m).

Example 2. The ring of rational integers is too special to give much insight into the primary-prime relationship. We shall now give an example of *a primary ideal which is not a power of the prime ideal to which it belongs.* In the polynomial ring $F[x, y]$, where F is a field, consider the ideal (x, y). A polynomial $g(x, y)$ belongs to (x, y) if, and only if, $g(0, 0)=0$, which shows immediately that (x, y) is a prime ideal. Again, if $f=f(x, y)$ does not belong to (x, y), then $f(0, 0)\neq 0$ so that $f(0, 0)$ is a unit of the ring. But $f(0, 0)$ belongs to $(f)+(x, y)$, consequently $(x, y)+(f)=(1)$, which shows that (x, y) is a *maximal prime ideal.* Next consider the ideal (x, y^2). We have $(x, y)^2=(x^2, xy, y^2)\subset(x, y^2)\subset(x, y)$, where the inclusions are strict. By Proposition 9, (x, y^2) is (x, y)-primary, but the inclusion relations just given show that (x, y^2) is not a power of (x, y).

Example 3. It is a less trivial matter to give *an example of a power of a prime ideal which is not primary,* for if we use, for this purpose, polynomials whose coefficients are taken from a field, then we must employ at least three variables. In the ring $F[x, y, z]$ (F is a field) we form an ideal $\mathfrak{p}$ by requiring that $f(x, y, z)$ shall belong to $\mathfrak{p}$ if, and only if, $f(t^3, t^4, t^5)$ be zero as a polynomial in t. It is clear that $\mathfrak{p}$ is prime and that $f_1=y^2-xz$, $f_2=yz-x^3$ and $f_3=z^2-x^2y$ are all in $\mathfrak{p}$. But f_1, f_2 and f_3 are not only in $\mathfrak{p}$, they actually generate $\mathfrak{p}$. We shall outline a proof of this and leave the reader to fill in the details. By means of the relations $f_1=y^2-xz$, $f_2=yz-x^3$ and $f_3=z^2-x^2y$ an arbitrary polynomial $f=f(x,y,z)$ can be written in the form

$$f(x, y, z)=x^2A(z)+xyB(z)+xC(z)+yD(z)+E(z)+g(x, y, z),$$

where $A(z), B(z), \dots, E(z)$ are polynomials in z alone, and where $g(x, y, z)$ belongs to (f_1, f_2, f_3). If now f belongs to $\mathfrak{p}$ we have

$$0=f(t^3, t^4, t^5)=t^6A(t^5)+t^7B(t^5)+t^3C(t^5)+t^4D(t^5)+E(t^5).$$

But here there can be no cancellation; for instance, a term of $t^6A(t^5)$ cannot have the same degree as a term of $t^7B(t^5)$. Thus $A(t^5), B(t^5), \dots, E(t^5)$ are all zero, and therefore $A(z)=B(z)=\dots=E(z)=0$; so that $f(x, y, z)=g(x, y, z)$ which belongs to (f_1, f_2, f_3).

We shall next show that $\mathfrak{p}^2$ is not primary. First, every prime ideal which contains $\mathfrak{p}^2$ must contain $\mathfrak{p}$, from which it follows that if $\mathfrak{p}^2$ were primary, it would be $\mathfrak{p}$-primary. Now $\mathfrak{p}^2$ contains

$$f_2^2-f_1f_3=x(x^5-3x^2yz+xy^3+z^3)$$

and x is not in $\mathfrak{p}$, hence if $\mathfrak{p}^2$ were $\mathfrak{p}$-primary $x^5-3x^2yz+xy^3+z^3$ would be in $\mathfrak{p}^2$. This, however, is not the case, for every polynomial in $\mathfrak{p}^2=(y^2-zx, yz-x^3, z^2-x^2y)^2$ contains no terms of degree less than four.

Example 4. Our remaining example is constructed to show that, in a normal decomposition, *the primary component belonging to an embedded prime ideal need not be unique.* To show this we consider the ideal $(x^2, xy)=(x)(x, y)$ in the ring $F[x, y]$, where F is a field, and show first that

$$(x^2, xy)=(x)\cap(y+ax, x^2), \tag{A}$$

where a is any element of F. It is clear that the right-hand side of (A) contains the left-hand side. Now suppose that $f=f(x, y)$ belongs to $(x)\cap(y+ax, x^2)$, then $f(x, y)=xg(x, y)=x(\alpha+\beta x+\gamma y+\ldots)$. But x^2 and xy are both in $(y+ax, x^2)$, consequently $x(\beta x+\gamma y+\ldots)$ is in $(y+ax, x^2)$ and therefore αx belongs to $(y+ax, x^2)$. However, αx can only be in $(y+ax, x^2)$ if $\alpha=0$, hence $f(x, y)=x(\beta x+\gamma y+\ldots)$, which shows that $f(x, y)$ belongs to (x^2, xy). The relation (A) has now been proved. It is clear that (x) is a prime ideal. Since

$$(x, y)^2=(x^2, xy, y^2)\subset(y+ax, x^2)\subset(x, y)$$

it follows that $(y+ax, x^2)$ is (x, y)-primary, because (x, y) is a maximal prime ideal (see Example 2). These results, when combined, show that (A) is a primary decomposition, and as neither of (x) and $(y+ax, x^2)$ contains the other, the decomposition is, in fact, normal. Hence, if a and a' are two *different* elements of F, then $(x^2, xy)=(x)\cap(y+ax, x^2)$ and $(x^2, xy)=(x)\cap(y+a'x, x^2)$ are normal decompositions in which the (x, y)-primary components are not the same.

It is worth noting that $(x^2, xy)=(x)\cap(x^2, xy, y^2)$ is yet another normal decomposition.

CHAPTER II

RESIDUE RINGS AND RINGS OF QUOTIENTS

2·1. Homomorphisms and isomorphisms. There are associated with a given ring R certain other rings, which can be derived from it by means of algebraic processes. The residue rings of R and the rings of quotients of R are among these derived rings, and they are of particular importance, because with their aid it is possible to simplify some of the problems which have to do with the ideals of R itself. In order to study the relation of R to one of its residue rings, we shall first define what we mean by a *homomorphic* mapping of R.

Suppose that we have a mapping, σ, of R into R', where R' may, at first, be a set of objects of any kind. Thus with each element $a \in R$ there is associated a definite element $\sigma(a)$ of R'. If it happens that every element of R' is the image of at least one element of R, then we say that σ maps R *on to* R'. Assume now that an addition and a multiplication have been defined on R'. By this we mean that if a' and b' belong to R', then $a'+b'$ and $a'b'$ have been defined and are again elements of R'; but it is not to be assumed that those relations hold,† which must be satisfied if R' is to be a ring. We can then state an important principle, namely:

If σ maps R on to R' in such a way that whenever a and b belong to R, then $\sigma(a+b)=\sigma(a)+\sigma(b)$ and $\sigma(ab)=\sigma(a)\,\sigma(b)$; and if, moreover, R' contains at least two elements, then R' is a ring with respect to the addition and multiplication which are defined on it. Further, $\sigma(0)$ and $\sigma(1)$ are the zero and unit elements of R', respectively.

To prove this we let a, b and c be elements of R, and put $\sigma(a)=a'$, $\sigma(b)=b'$ and $\sigma(c)=c'$. Applying σ to both sides of the equation $a+b=b+a$ we obtain

$$\sigma(a)+\sigma(b)=\sigma(a+b)=\sigma(b+a)=\sigma(b)+\sigma(a),$$

that is, $a'+b'=b'+a'$. Similarly, from $a+(b+c)=(a+b)+c$,

† Such as $a'+b'=b'+a'$.

from $ab=ba$, from $a(bc)=(ab)c$ and from $a(b+c)=ab+ac$ are obtained

$$a'+(b'+c')=(a'+b')+c', \quad a'b'=b'a', \quad a'(b'c')=(a'b')c'$$

and $a'(b'+c')=a'b'+a'c'$. Again, from $a+0=a$ and from $a+(-a)=0$ it follows that $a'+\sigma(0)=a'$, and that

$$a'+\sigma(-a)=\sigma(0).$$

By hypothesis, σ maps R *on to* R', hence a', b', c' can be any three elements of R', and accordingly R' is a commutative ring in the abstract sense (see Preliminaries) having $\sigma(0)$ as its zero element. All that remains to be proved is that $\sigma(1)\,a'=a'$ and that $\sigma(1)\neq\sigma(0)$. The first of these assertions follows from the equation $1a=a$. Let us assume that $\sigma(1)=\sigma(0)$ and see if we obtain a contradiction. By our assumption, $a'=\sigma(1)\,a'=\sigma(0)\,a'$, consequently $a'=\sigma(0)$, because we have shown that $\sigma(0)$ is the zero element of R'. Thus R' consists of the single element $\sigma(0)$, and this is a contradiction, for it is given that R' contains at least two elements. The proof is now complete.

DEFINITION. *If a mapping σ of a ring R on to a ring R' is such that $\sigma(a+b)=\sigma(a)+\sigma(b)$ and $\sigma(ab)=\sigma(a)\,\sigma(b)$ for all pairs a, b of elements of R, then we say that 'σ is a homomorphism of R on to R''.*

Suppose that σ maps R homomorphically on to R', then, since $a+(-a)=0$ and since $\sigma(0)$ is the zero element of R', it follows that $\sigma(-a)=-\sigma(a)$.

We can, of course, have a homomorphism σ of a ring R *into* a ring R'. In this case, the reader should note that although $\sigma(0)$ must be the zero element of R', it can easily happen that $\sigma(1)$ is not the unit element of R'.

DEFINITION. *If a homomorphism σ of R on to R' is such that σ sets up a* 1-1 *correspondence between the elements of R and those of R', then we say that 'σ maps R isomorphically on to R'', and we say, also, that 'R and R' are isomorphic'.*

If σ is an isomorphism of R on to R', then it is clear that the inverse mapping, σ^{-1}, will map R' isomorphically on to R. Two rings which are isomorphic are faithful copies of each other, and, in consequence, they have the same algebraic properties.

2·2. Residue rings. Let $\mathfrak{a}$ be a proper ideal of the ring R, and suppose that x_1 and x_2 are two elements of R. If $x_1 - x_2 \in \mathfrak{a}$, we say that x_2 *is congruent to* x_1 *modulo* $\mathfrak{a}$, and we write either $x_1 \equiv x_2 \pmod{\mathfrak{a}}$, or $x_1 \equiv x_2(\mathfrak{a})$. This relationship between elements is reflexive, symmetric, and transitive. By saying that it is *reflexive* we mean that $x \equiv x \pmod{\mathfrak{a}}$ for all $x \in R$; by *symmetric* we mean that if $x \equiv y \pmod{\mathfrak{a}}$ then $y \equiv x \pmod{\mathfrak{a}}$; and by *transitive* we mean that $x \equiv y \pmod{\mathfrak{a}}$ and $y \equiv z \pmod{\mathfrak{a}}$ together imply that $x \equiv z \pmod{\mathfrak{a}}$. Let us now collect the elements of R into classes of mutually congruent elements, so that two elements in the same class will be congruent; but if we first take an element from one class and afterwards take an element from a different class, then these two elements will not be congruent. The classes of elements are known as the *residue classes* of $\mathfrak{a}$. Since we are supposing that $\mathfrak{a}$ is a proper ideal, 1 and 0 cannot be in the same residue class, hence there exist at least two residue classes. We propose, with the aid of the following lemma, to make a ring out of the residue classes.

LEMMA 1. *If* $x_1 \equiv x_2 \pmod{\mathfrak{a}}$ *and if* $y_1 \equiv y_2 \pmod{\mathfrak{a}}$, *then modulo* $\mathfrak{a}$ *we have* $x_1 + y_1 \equiv x_2 + y_2$, $x_1 - y_1 \equiv x_2 - y_2$, *and* $x_1 y_1 \equiv x_2 y_2$.

Proof.
$$(x_1 + y_1) - (x_2 + y_2) = (x_1 - x_2) + (y_1 - y_2),$$
$$(x_1 - y_1) - (x_2 - y_2) = (x_1 - x_2) - (y_1 - y_2)$$
and
$$x_1 y_1 - x_2 y_2 = (x_1 - x_2)\, y_1 + x_2(y_1 - y_2).$$

But $x_1 - x_2$, $y_1 - y_2$, $(x_1 - x_2)\, y_1$ and $x_2(y_1 - y_2)$ are all in $\mathfrak{a}$, hence $(x_1 + y_1) - (x_2 + y_2)$, $(x_1 - y_1) - (x_2 - y_2)$ and $x_1 y_1 - x_2 y_2$ are all in $\mathfrak{a}$. This is what we had to prove.

Now let X and Y be two residue classes of $\mathfrak{a}$, let x_1, x_2 be elements of X, and let y_1, y_2 be elements of Y. By Lemma 1, $x_1 + y_1 \equiv x_2 + y_2$, so that the residue class of $x_1 + y_1$ depends only on X and Y, and not on the particular representatives x_1 of X and y_1 of Y, which we used to obtain it. This residue class is called the *sum* of X and Y, and it is denoted by $X + Y$. Similarly, the residue class of $x_1 y_1$ depends only on X and Y; it is called the *product* of X and Y, and it is denoted by XY. The residue classes now form a ring, for if σ is the mapping which sends each element x into its residue class X, then, by our construction,

$$\sigma(x'+x'')=X'+X'' \quad \text{and} \quad \sigma(x'x'')=X'X'',$$

so that our assertion follows from the general principle enunciated in §2·1. The ring of residue classes is known as the *residue class ring of R modulo* $\mathfrak{a}$, and it is denoted by $R/\mathfrak{a}$. Our results include

PROPOSITION 1. *The mapping which maps each element into its residue class modulo* $\mathfrak{a}$, *is a homomorphism.*

It will be called the *natural homomorphism* of R on to $R/\mathfrak{a}$.

PROPOSITION 2. *Let* σ *be a homomorphism of a ring* R *on to a ring* R', *and let* $\mathfrak{n}$ *consist of all those elements of* R *which are mapped into the zero element of* R'. *Then* $\mathfrak{n}$ *is a proper ideal of* R, *and we have* $\sigma(x)=\sigma(y)$ *if, and only if,* $x \equiv y \pmod{\mathfrak{n}}$. *Thus* σ *determines a mapping of the residue classes of* $\mathfrak{n}$, *and this mapping is an isomorphism of* $R/\mathfrak{n}$ *on to* R'.

Proof. Let x_1, x_2 belong to $\mathfrak{n}$, and let $r \in R$. Then

$$\sigma(x_1 \pm x_2)=\sigma(x_1) \pm \sigma(x_2)=0 \pm 0=0,$$

and also
$$\sigma(rx_1)=\sigma(r)\,\sigma(x_1)=\sigma(r)\,0=0,$$

which shows that $\mathfrak{n}$ is an ideal. Now $\sigma(1)$ is the unit element of R', hence $1 \notin \mathfrak{n}$, so that $\mathfrak{n}$ is a proper ideal. Again, $\sigma(x)=\sigma(y)$ if, and only if, $\sigma(x-y)=0$; hence $\sigma(x)=\sigma(y)$ if, and only if, $x-y \in \mathfrak{n}$. The second assertion of the proposition has now been proved, and it shows that all the elements in a single residue class of $\mathfrak{n}$ are mapped by σ into the same element. Thus σ determines a mapping of the residue classes, or, in other words, σ determines a mapping of $R/\mathfrak{n}$. The remaining assertions are now obvious.

DEFINITION. *If* σ *is a homomorphism of a ring* R *on to a ring* R', *the ideal* $\mathfrak{n}$, *which consists of all the elements of* R *mapped by* σ *into the zero element of* R', *is called the 'nucleus' or 'kernel' of the mapping. The mapping* σ *is an isomorphism if, and only if,* $\mathfrak{n}=(0)$.

Suppose that σ is a mapping of R into R', that A is a set of elements of R and that A' is a set of elements of R'. We then define sets $\sigma(A)$ and $\sigma^{-1}(A')$ as follows: $\sigma(A)$ shall consist of all elements of the form $\sigma(a)$, where $a \in A$; and $\sigma^{-1}(A')$ shall consist of every element of R which has its image in A'.

PROPOSITION 3. *Let σ be a homomorphism of R on to R', and let $\mathfrak{n}$ be its nucleus. Then there is a 1-1 correspondence between the ideals $\mathfrak{a}$ of R, which contain $\mathfrak{n}$, and the ideals $\mathfrak{a}'$ of R', such that if $\mathfrak{a}$ and $\mathfrak{a}'$ correspond then $\sigma(\mathfrak{a}) = \mathfrak{a}'$ and $\mathfrak{a} = \sigma^{-1}(\mathfrak{a}')$.*

Proof. Let $\mathfrak{a}'$ be an ideal of R', then it is a trivial matter to verify that $\sigma^{-1}(\mathfrak{a}')$ is an ideal of R which contains $\mathfrak{n}$. Further, since σ maps R on to R', we have $\sigma(\sigma^{-1}(\mathfrak{a}')) = \mathfrak{a}'$, and hence, in particular, if $\mathfrak{a}_1'$ and $\mathfrak{a}_2'$ are different ideals of R' then $\sigma^{-1}(\mathfrak{a}_1')$ and $\sigma^{-1}(\mathfrak{a}_2')$ are different. Let $\mathfrak{a}$ be an ideal of R containing $\mathfrak{n}$. We shall show that $\mathfrak{a} = \sigma^{-1}(\mathfrak{a}')$ where $\mathfrak{a}'$ is a suitable ideal of R', and this will complete the proof. Put $\mathfrak{a}' = \sigma(\mathfrak{a})$, then, as may be verified immediately, $\mathfrak{a}'$ is an ideal and $\mathfrak{a} \subseteq \sigma^{-1}(\mathfrak{a}')$. Next let $x \in \sigma^{-1}(\mathfrak{a}')$, then $\sigma(x) \in \mathfrak{a}' = \sigma(\mathfrak{a})$, hence we can find $a \in \mathfrak{a}$ such that $\sigma(x) = \sigma(a)$. By Proposition 2, $x - a \in \mathfrak{n} \subseteq \mathfrak{a}$ which shows that $x \in \mathfrak{a}$. It has now been proved that $\sigma^{-1}(\mathfrak{a}') \subseteq \mathfrak{a}$, and this, combined with our other results, establishes the proposition.

In the corollaries which follow, we suppose that we have the situation described in the statement of Proposition 3, and when we say that $\mathfrak{a}$ and $\mathfrak{a}'$ are corresponding ideals, we mean that $\sigma(\mathfrak{a}) = \mathfrak{a}'$ and that $\mathfrak{a} = \sigma^{-1}(\mathfrak{a}')$.

COROLLARY 1. *Let $\mathfrak{a}$ and $\mathfrak{a}'$ be corresponding ideals. If one of them is proper so is the other, and then $R/\mathfrak{a}$ and $R'/\mathfrak{a}'$ are isomorphic.*

Proof. If $\mathfrak{a} = R$ then $\mathfrak{a}' = \sigma(R) = R'$, while if $\mathfrak{a}' = R'$ then $\mathfrak{a} = \sigma^{-1}(R') = R$. This establishes the first point. Now suppose that $\mathfrak{a}$ and $\mathfrak{a}'$ are proper ideals and let τ be the natural homomorphism of R' on to $R'/\mathfrak{a}'$. The mapping $x \to \tau(\sigma(x))$ is a homomorphism of R on to $R'/\mathfrak{a}'$, and x is mapped into zero if, and only if, $\sigma(x) \in \mathfrak{a}'$; i.e. if and only if $x \in \sigma^{-1}(\mathfrak{a}') = \mathfrak{a}$. By Proposition 2, it now follows that $R/\mathfrak{a}$ and $R'/\mathfrak{a}'$ are isomorphic.

COROLLARY 2. *Suppose that $\mathfrak{a}_i$ corresponds to $\mathfrak{a}_i'$ for each i in a set I. Then $\bigcap_{i \in I} \mathfrak{a}_i$ corresponds to $\bigcap_{i \in I} \mathfrak{a}_i'$.*

This follows from the obvious relation

$$\sigma^{-1}(\bigcap \mathfrak{a}_i') = \bigcap \sigma^{-1}(\mathfrak{a}_i').$$

2·3. Further properties of residue rings. In this section we shall obtain some results which connect the theory of residue rings with the theory developed in Chapter I. First, however, we require a definition.

DEFINITION. *An element x of a ring R is said to be 'nilpotent' if some positive power of x is zero.*

PROPOSITION 4. *Let $\mathfrak{a}$ be a proper ideal of R, then*

(1) *$\mathfrak{a}$ is a prime ideal if, and only if, $R/\mathfrak{a}$ is an integral domain;*

(2) *$\mathfrak{a}$ is a primary ideal if, and only if, every zero divisor in $R/\mathfrak{a}$ is nilpotent.*

Proof. These results follow easily from the definitions, so, as an example, we shall prove (2) and we shall leave the proof of (1) to the reader. If $x \in R$ we shall use $\bar{x}$ to denote its residue class modulo $\mathfrak{a}$. First, suppose that $\mathfrak{a}$ is primary and that $\bar{x}$ is a zero divisor in $R/\mathfrak{a}$. Then we can find $y \in R$ such that $\bar{y} \neq 0$ and $\bar{x}\bar{y} = 0$. Thus $\overline{xy} = \bar{x}\bar{y} = 0$ so that $xy \in \mathfrak{a}$, and we also have $y \notin \mathfrak{a}$ since $\bar{y} \neq 0$. But $\mathfrak{a}$ is primary, consequently there is an integer n such that $x^n \in \mathfrak{a}$, from which it follows that $\bar{x}^n = 0$ so that $\bar{x}$ is nilpotent. Secondly, suppose that in $R/\mathfrak{a}$ every zero divisor is nilpotent, and assume that $xy \in \mathfrak{a}$ where $y \notin \mathfrak{a}$. Then $\bar{x}\bar{y} = 0$ and $\bar{y} \neq 0$, hence $\bar{x}$ is a zero divisor. By hypothesis, there exists an integer n such that $\bar{x}^n = 0$, which shows that $x^n \in \mathfrak{a}$. Thus $\mathfrak{a}$ is primary.

In Proposition 5 and its corollary we suppose that we have a homomorphism σ of a ring R on to a ring R', and, as in § 2·2, we shall say that an ideal $\mathfrak{a}$ of R corresponds to an ideal $\mathfrak{a}'$ of R', if $\sigma(\mathfrak{a}) = \mathfrak{a}'$ and $\sigma^{-1}(\mathfrak{a}') = \mathfrak{a}$.

PROPOSITION 5. *Let $\mathfrak{a}$ and $\mathfrak{a}'$ be corresponding proper ideals, then if one is primary so is the other, and if one is prime so is the other.*

Proof. By Corollary 1 of Proposition 3, the rings $R/\mathfrak{a}$ and $R'/\mathfrak{a}'$ are isomorphic. If, therefore, one of them has the property that every zero divisor is nilpotent, then the other one will have the same property. Again, if one of these rings is an integral domain so is the other. Proposition 5 now follows from Proposition 4.

COROLLARY 1. *Suppose that* $\mathfrak{q}$ *and* $\mathfrak{q}'$ *are corresponding primary ideals, and that* $\mathfrak{p}$ *and* $\mathfrak{p}'$ *are corresponding prime ideals. Then* $\mathfrak{q}$ *is* $\mathfrak{p}$*-primary if, and only if,* $\mathfrak{q}'$ *is* $\mathfrak{p}'$*-primary.*

Proof. It is clear that $\mathfrak{p}$ will be a minimal prime ideal of $\mathfrak{q}$ if, and only if, $\mathfrak{p}'$ is a minimal prime ideal of $\mathfrak{q}'$.

It is convenient, at this stage, to make some remarks about the maximal ideals of a ring. The maximal ideals are defined thus:

DEFINITION. *A proper ideal* $\mathfrak{a}$ *of a ring* R *is called a 'maximal ideal' of* R*, if there is no other proper ideal which contains* $\mathfrak{a}$.

The maximal ideals of R may be characterized in the following way:

PROPOSITION 6. *Let* $\mathfrak{a}$ *be a proper ideal of* R*, then* $R/\mathfrak{a}$ *is a field if, and only if,* $\mathfrak{a}$ *is a maximal ideal.*

Proof. By Proposition 10 of § 1·10, $R/\mathfrak{a}$ is a field if and only if (0) and (1) are its only ideals. This, however, occurs when and only when there is no ideal strictly between $\mathfrak{a}$ and the whole ring R.

Since a field is always an integral domain, Propositions 4 and 6 show that a maximal ideal is necessarily a prime ideal, consequently a maximal ideal is always a maximal prime ideal *in the sense of* § 1·9. It is also true that every maximal prime ideal, in the sense of § 1·9, is a maximal ideal and to prove this it would be sufficient to show that every proper ideal $\mathfrak{b}$ is contained in at least one maximal ideal. In the case of a Noetherian ring this latter assertion follows at once, for the set $\mathfrak{S}$ of all proper ideals containing $\mathfrak{b}$ will contain one that is maximal for $\mathfrak{S}$, and this ideal will be a maximal ideal of R. In the case of an arbitrary ring the required result can be established by transfinite methods, but we shall not stop to show this here. Let us note, however, that our observations have given us a proof of the following: *If* $\mathfrak{p}$ *is a maximal prime ideal of a Noetherian ring* R*, then* $R/\mathfrak{p}$ *is a field.*

We end this section by showing that all the residue rings of a Noetherian ring are themselves Noetherian.

PROPOSITION 7. *A homomorphic image of a Noetherian ring is again Noetherian.*

Proof. Let σ be a homomorphism of a Noetherian ring R on to a ring R', and let $\mathfrak{a}'$ be an ideal of R'. Put $\mathfrak{a}=\sigma^{-1}(\mathfrak{a}')$, then $\mathfrak{a}$ is an ideal of R and is therefore finitely generated, say

$$\mathfrak{a}=(a_1, a_2, \ldots, a_n).$$

But $\sigma(a_1),\sigma(a_2),\ldots,\sigma(a_n)$ will generate $\sigma(\mathfrak{a})=\mathfrak{a}'$, hence $\mathfrak{a}'$ is finitely generated.

2·4. Extension rings. Let R and R' be two rings, and suppose that every element of R is also an element of R', then if a and b belong to R they will have a sum according to the addition in R, which we will denote by $(a+b)_R$; and they will also have a sum according to the addition in R', which we will denote by $(a+b)_{R'}$. In a similar manner we define $(ab)_R$ and $(ab)_{R'}$. *If it should happen that* $(a+b)_R=(a+b)_{R'}$ *and* $(ab)_R=(ab)_{R'}$, *for all pairs a, b of elements of R; and if, moreover, the unit element of R be the same as the unit element of R', then we say that 'R' is an extension ring of R'*. When R' is an extension ring of R, the zero element O_R of R is the same as the zero element $O_{R'}$ of R'. For $O_R+O_R=O_R$ is an equation which holds in R, and therefore it also holds in R'; but in R' we have $O_R+O_{R'}=O_R$ hence, by the uniqueness of subtraction in R', $O_R=O_{R'}$.

Suppose that R' is an extension ring of R and let $\mathfrak{a}$ be an ideal of R. The ideal generated by $\mathfrak{a}$ in R' (see §1·3), that is, the set of all finite sums $r'_1a_1+r'_2a_2+\ldots+r'_na_n$, where $r'_i\in R'$ and $a_i\in\mathfrak{a}$, is called *the extension of* $\mathfrak{a}$ *to* R', and it is denoted by $R'\mathfrak{a}$. Again, if $\mathfrak{a}'$ is an ideal of R' then *by the contraction of* $\mathfrak{a}'$ *in* R we mean $\mathfrak{a}'\cap R$, which is clearly an ideal of R. We shall not consider the operations of extension and contraction in the general situation, apart from drawing attention to the following elementary, but important, result.

PROPOSITION 8. *Let R' be an extension ring of R, let $\mathfrak{p}'$ be a prime ideal of R', and let $\mathfrak{q}'$ be $\mathfrak{p}'$-primary. Put $\mathfrak{q}=\mathfrak{q}'\cap R$ and $\mathfrak{p}=\mathfrak{p}'\cap R$, then $\mathfrak{p}$ is prime and $\mathfrak{q}$ is $\mathfrak{p}$-primary.*

The proof is a simple and straightforward application of Lemma 1 of § 1·5.

2·5. The full ring of quotients. Let R be a given ring, and let $\mathfrak{S}$ be the set of all elements which are not zero divisors, so that in particular, $1 \in \mathfrak{S}$. The set $\mathfrak{S}$ is multiplicatively closed. For suppose that $c_1 \in \mathfrak{S}$, that $c_2 \in \mathfrak{S}$, and that $c_1 c_2 x = 0$, then $c_1(c_2 x) = 0$ and c_1 is not a zero divisor, consequently $c_2 x = 0$ and hence $x = 0$. Thus $c_1 c_2$ is not a zero divisor, and, accordingly, $c_1 c_2 \in \mathfrak{S}$. We now consider ordered pairs of elements in which the first constituent is an arbitrary element a of R, and in which the second constituent is any element c of $\mathfrak{S}$. For reasons which will become clear later, it is convenient to write the ordered pair as $\frac{a}{c}$, and to refer to it as a *formal quotient*. If $\frac{a_1}{c_1}$ and $\frac{a_2}{c_2}$ are two formal quotients, let us write $\frac{a_1}{c_1} \sim \frac{a_2}{c_2}$ and say that $\frac{a_1}{c_1}$ and $\frac{a_2}{c_2}$ are *equivalent* if $a_1 c_2 = a_2 c_1$. We assert that:

(1) $\frac{a}{c} \sim \frac{a}{c}$;

(2) if $\frac{a_1}{c_1} \sim \frac{a_2}{c_2}$ then $\frac{a_2}{c_2} \sim \frac{a_1}{c_1}$;

(3) if $\frac{a_1}{c_1} \sim \frac{a_2}{c_2}$ and $\frac{a_2}{c_2} \sim \frac{a_3}{c_3}$ then $\frac{a_1}{c_1} \sim \frac{a_3}{c_3}$.

Of these (1) and (2) are trivial. In (3) we are given that $a_1 c_2 = a_2 c_1$ and $a_2 c_3 = a_3 c_2$, consequently $a_1 c_2 c_3 = a_2 c_1 c_3 = a_3 c_2 c_1$. But c_2 *is not a zero divisor*, hence from $a_1 c_2 c_3 = a_3 c_2 c_1$ we obtain $a_1 c_3 = a_3 c_1$ that is, $\frac{a_1}{c_1} \sim \frac{a_3}{c_3}$. It follows from (1), (2) and (3) that the formal quotients can be grouped into classes which have the following properties: no two classes have a common member, and two formal quotients are equivalent if, and only if, they belong to the same class. As a temporary device, we shall use $\left\{\frac{a}{c}\right\}$ to denote the class to which $\frac{a}{c}$ belongs.

LEMMA 2. *If* $\dfrac{a_1}{c_1} \sim \dfrac{a_1'}{c_1'}$ *and* $\dfrac{a_2}{c_2} \sim \dfrac{a_2'}{c_2'}$ *then* $\dfrac{a_1c_2+a_2c_1}{c_1c_2} \sim \dfrac{a_1'c_2'+a_2'c_1'}{c_1'c_2'}$ *and* $\dfrac{a_1a_2}{c_1c_2} \sim \dfrac{a_1'a_2'}{c_1'c_2'}$.

Proof. It is given that $a_1c_1' = a_1'c_1$ and that $a_2c_2' = a_2'c_2$. From these it follows at once that $(a_1c_2+a_2c_1)c_1'c_2' = (a_1'c_2'+a_2'c_1')c_1c_2$ and that $a_1a_2c_1'c_2' = a_1'a_2'c_1c_2$, which are precisely the relations that we have to prove.

Now let A_1 and A_2 be two classes of formal quotients, let $\dfrac{a_1}{c_1}, \dfrac{a_1'}{c_1'}$ belong to A_1, and let $\dfrac{a_2}{c_2}, \dfrac{a_2'}{c_2'}$ belong to A_2. By Lemma 2, $\dfrac{a_1c_2+a_2c_1}{c_1c_2}$ and $\dfrac{a_1'c_2'+a_2'c_1'}{c_1'c_2'}$ belong to the same class, so that the class of $\dfrac{a_1c_2+a_2c_1}{c_1c_2}$ depends only on A_1 and A_2, and not on the particular representatives $\dfrac{a_1}{c_1}$ (of A_1) and $\dfrac{a_2}{c_2}$ (of A_2) which were used to construct it. This class will be called the *sum* of A_1 and A_2 and it will be denoted by $A_1 + A_2$. A similar argument shows that the class of $\dfrac{a_1a_2}{c_1c_2}$ depends only on A_1 and A_2; it is called the *product* of A_1 and A_2 and it is denoted by A_1A_2. With these definitions of addition and multiplication the classes of formal quotients form a commutative ring, which has $\left\{\dfrac{0}{1}\right\}$ as its zero element and $\left\{\dfrac{1}{1}\right\}$ as its unit element. A full justification of this statement would require several simple but tedious verifications, so we shall leave the reader to make these verifications at his leisure. The ring, which has just been constructed, is called *the full ring of quotients of R.*

The rather cumbersome notation, which was introduced to explain the construction, can now be simplified. We shall let a formal quotient $\dfrac{a}{c}$ stand both for itself and for the class which it determines, so that instead of writing

$$\left\{\frac{a_1}{c_1}\right\} = \left\{\frac{a_1'}{c_1'}\right\}, \quad \left\{\frac{a_1}{c_1}\right\} + \left\{\frac{a_2}{c_2}\right\} = \left\{\frac{a_1c_2+a_2c_1}{c_1c_2}\right\},$$

and so on, we shall write $\frac{a_1}{c_1}=\frac{a_1'}{c_1'}$, $\frac{a_1}{c_1}+\frac{a_2}{c_2}=\frac{a_1c_2+a_2c_1}{c_1c_2}$, and so on. Again, the mapping $a\to\frac{a}{1}$ is clearly an isomorphism of R into the full ring of quotients. In future, we shall use a to denote both itself and the class $\frac{a}{1}$, and this, in practice, causes no confusion, because either the context makes clear which interpretation is intended, or the various plausible interpretations are all equivalent. It is usual to describe what has just been done by saying that *we have identified R with the ring which consists of all classes of the form* $\frac{a}{1}$. Since the full ring of quotients is an extension ring of the one with which R has been identified, we say that *after identification, the full ring of quotients is an extension ring of R.* Finally, we shall often use a/c instead of $\frac{a}{c}$; the reasons for this are entirely typographical.

If R is an integral domain, the set $\mathfrak{S}$ of elements which are not zero divisors consists of all the elements which are not zero. Let a/c be a non-zero element of the full ring $\mathfrak{R}$ of quotients, so that $a\in R$, $c\in R$, $a\neq 0$, and $c\neq 0$. Then a/c has an inverse in $\mathfrak{R}$, namely, c/a, consequently $\mathfrak{R}$ is a field. We call this field the *quotient field* of R. For example, if R is the ring of positive and negative integers, then its quotient field is the field of rational numbers.

2·6. Rings of quotients. Let R be a ring and let $\mathfrak{R}$ be its full ring of quotients. Throughout § 2·6 S will denote a multiplicatively closed set of elements of R *none of which is a zero divisor in* R; but in § 2·7 we shall weaken this assumption and at the same time extend most of the results which we are about to prove. The set of all elements of $\mathfrak{R}$ which can be written in the form a/c, where $a\in R$ and $c\in S$, is an extension ring of R. This extension ring is denoted by R_S and called *the ring of quotients of R, formed with respect to S.*

PROPOSITION 9. If $\mathfrak{a}'$ *is an ideal of* $R'=R_S$, *then* $R'(\mathfrak{a}'\cap R)=\mathfrak{a}'$.

Proof. It is enough to prove that $\mathfrak{a}' \subseteq R'(\mathfrak{a}' \cap R)$. Let $x \in \mathfrak{a}'$ then $x = r/c$, where $r \in R$ and $c \in S$, hence $r = cx \in (\mathfrak{a}' \cap R)$ and, accordingly, $x = \frac{1}{c} r$ belongs to $R'(\mathfrak{a}' \cap R)$. This shows that $\mathfrak{a}' \subseteq R'(\mathfrak{a}' \cap R)$.

PROPOSITION 10. *If R is Noetherian then so is $R' = R_S$.*

Proof. Let $\mathfrak{a}'$ be an ideal of R', then $\mathfrak{a}' \cap R$ is an ideal of R and therefore it is finitely generated, say

$$\mathfrak{a}' \cap R = Ra_1 + Ra_2 + \ldots + Ra_n.$$

By Proposition 9,

$$\begin{aligned}\mathfrak{a}' = R'(\mathfrak{a}' \cap R) = R'(Ra_1 + Ra_2 + \ldots + Ra_n)\\ = R'a_1 + R'a_2 + \ldots + R'a_n,\end{aligned}$$

and therefore $\mathfrak{a}'$ is finitely generated.

LEMMA 3. *Let $\mathfrak{a}$ be an ideal of R and let $R' = R_S$, then an element x (of the full ring of quotients of R) belongs to $R'\mathfrak{a}$ if, and only if, it can be written in the form $x = a/c$, where $a \in \mathfrak{a}$ and $c \in S$.*

Proof. Let $x \in R'\mathfrak{a}$, then x can be written in the form

$$x = r_1' a_1 + r_2' a_2 + \ldots + r_n' a_n,$$

where $r_i' \in R'$ and $a_i \in \mathfrak{a}$. Since $r_i' \in R'$, $r_i' = r_i/c_i$ where $r_i \in R$ and $c_i \in S$. Put $c_1 c_2 \ldots c_n = c$, then with suitable elements $\rho_1, \rho_2, \ldots, \rho_n$ of R we have $x_i' = \rho_i/c$, and therefore

$$x = (\rho_1 a_1 + \rho_2 a_2 + \ldots + \rho_n a_n)/c = a/c,$$

where $a \in \mathfrak{a}$. This proves half of the lemma, and the remaining half is obvious.

PROPOSITION 11. *Let $\mathfrak{p}$ be a prime ideal of R not meeting S, let $\mathfrak{q}$ be $\mathfrak{p}$-primary, and let $R' = R_S$. Then $\mathfrak{p}' = R'\mathfrak{p}$ is a prime ideal of R' and $\mathfrak{q}' = R'\mathfrak{q}$ is $\mathfrak{p}'$-primary. Further, $\mathfrak{p}' \cap R = \mathfrak{p}$ and $\mathfrak{q}' \cap R = \mathfrak{q}$.*

Proof. We shall show that $\mathfrak{q}'$ is $\mathfrak{p}'$-primary by using Lemma 1 of §1·5. It is clear that $\mathfrak{p}' \supseteq \mathfrak{q}'$. Let $x \in \mathfrak{p}'$, then, by Lemma 3, $x = p/c$, where $p \in \mathfrak{p}$ and $c \in S$. But $\mathfrak{q}$ is $\mathfrak{p}$-primary, hence we can choose an integer n such that $p^n \in \mathfrak{q}$, and then $x^n = p^n/c^n \in \mathfrak{q}'$. Now suppose that (with the obvious notations) $\alpha = r_1/c_1$ and $\beta = r_2/c_2$ are two elements of R' with the properties that $\alpha\beta \in \mathfrak{q}'$ and $\alpha \notin \mathfrak{p}'$.

Then $r_1 \notin \mathfrak{p}$ and, by Lemma 3, $\alpha\beta = r_1 r_2/c_1 c_2 = q/c$, where $q \in \mathfrak{q}$ and $c \in S$, consequently $c r_1 r_2 \in \mathfrak{q}$, which shows that $r_2 \in \mathfrak{q}$, because neither r_1 nor c is in $\mathfrak{p}$. Thus $\beta = r_2/c_2$ belongs to $\mathfrak{q}'$. We are now in a position to apply Lemma 1 of § 1·5, and it follows from that lemma that $\mathfrak{p}'$ is prime and that $\mathfrak{q}'$ is $\mathfrak{p}'$-primary.

In order to prove that $\mathfrak{q}' \cap R = \mathfrak{q}$ it will be enough to show that $\mathfrak{q}' \cap R \subseteq \mathfrak{q}$. Let $z \in \mathfrak{q}' \cap R$, then $z = q/c$, where $q \in \mathfrak{q}$ and $c \in S$, hence $cz \in \mathfrak{q}$ and $c \notin \mathfrak{p}$, consequently $z \in \mathfrak{q}$. Thus $\mathfrak{q}' \cap R \subseteq \mathfrak{q}$ and accordingly $\mathfrak{q}' \cap R = \mathfrak{q}$. By putting $\mathfrak{q} = \mathfrak{p}$ we obtain $\mathfrak{p}' \cap R = \mathfrak{p}$, and this completes the proof.

COROLLARY 1. *There is a* 1-1 *correspondence between the prime ideals* $\mathfrak{p}$, *which do not meet* S, *and the proper prime ideals* $\mathfrak{p}'$ *of* $R' = R_S$, *such that if* $\mathfrak{p}$ *and* $\mathfrak{p}'$ *correspond then* $\mathfrak{p}' = R'\mathfrak{p}$ *and* $\mathfrak{p} = \mathfrak{p}' \cap R$.

Proof. Let $\mathfrak{p}_1$ and $\mathfrak{p}_2$ be different prime ideals of R neither of which meets S, then by the proposition $\mathfrak{p}_1' = R'\mathfrak{p}_1$ and $\mathfrak{p}_2' = R'\mathfrak{p}_2$ are prime ideals of R'. Further, $\mathfrak{p}_1' \cap R = \mathfrak{p}_1$ and $\mathfrak{p}_2' \cap R = \mathfrak{p}_2$, which shows that $\mathfrak{p}_1'$ and $\mathfrak{p}_2'$ are proper, and that $\mathfrak{p}_1' \neq \mathfrak{p}_2'$. Finally, let $\mathfrak{p}'$ be a proper prime ideal of R' then $\mathfrak{p}'$ does not meet S (otherwise $\mathfrak{p}'$ would contain the unit element), and therefore $\mathfrak{p} = R \cap \mathfrak{p}'$ does not meet S. By Proposition 8, $\mathfrak{p}$ is a prime ideal, and by Proposition 9, $\mathfrak{p}' = R'\mathfrak{p}$. This completes the proof.

COROLLARY 2. *Let* $\mathfrak{p}$ *and* $\mathfrak{p}'$ *be corresponding prime ideals* (*in the sense of Corollary* 1) *of* R *and* $R' = R_S$, *then there is a* 1-1 *correspondence between the* $\mathfrak{p}$*-primary ideals* $\mathfrak{q}$ *and the* $\mathfrak{p}'$*-primary ideals* $\mathfrak{q}'$, *such that if* $\mathfrak{q}$ *and* $\mathfrak{q}'$ *correspond then* $\mathfrak{q}' = R'\mathfrak{q}$ *and* $\mathfrak{q} = R \cap \mathfrak{q}'$.

Proof. Let $\mathfrak{q}_1$ and $\mathfrak{q}_2$ be different $\mathfrak{p}$-primary ideals, then, by the proposition, $\mathfrak{q}_1' = R'\mathfrak{q}_1$ and $\mathfrak{q}_2' = R'\mathfrak{q}_2$ are both $\mathfrak{p}'$-primary; and, also, $\mathfrak{q}_1' \cap R = \mathfrak{q}_1$, $\mathfrak{q}_2' \cap R = \mathfrak{q}_2$. This shows, in particular, that $\mathfrak{q}_1' \neq \mathfrak{q}_2'$. Again, if $\mathfrak{q}'$ is $\mathfrak{p}'$-primary then, by Proposition 8, $\mathfrak{q} = \mathfrak{q}' \cap R$ is $\mathfrak{p}$-primary and, by Proposition 9, $\mathfrak{q}' = R'\mathfrak{q}$. This completes the proof.

Let us note that if $\mathfrak{q}$ is $\mathfrak{p}$-primary and $\mathfrak{p}$ *meets* S, say $c \in \mathfrak{p} \cap S$, then a high power of c will belong both to $\mathfrak{q}$ and to S, consequently $\mathfrak{q}$ also meets S. It follows that $R'\mathfrak{p}$ and $R'\mathfrak{q}$ both contain the unit element, hence $R'\mathfrak{p} = R'\mathfrak{q} = R'$. Thus we may say that *if* $\mathfrak{p}$ *meets* S, *every* $\mathfrak{p}$*-primary ideal is lost in* R_S.

2·7. Generalized rings of quotients. The investigations of the preceding section were made under the assumptions that S was a multiplicatively closed set of elements of R, and that no element of S was a zero divisor. We shall now weaken these assumptions by replacing them by: *S is multiplicatively closed and* $0 \notin S$. Let $\mathfrak{n}$ be the S-component of the ideal (0) (see § 1·7) and let σ be the natural homomorphism of R on to $R^* = R/\mathfrak{n}$, then $S^* = \sigma(S)$ will be a multiplicatively closed set in the ring R^*. R^*, S^*, σ and $\mathfrak{n}$ will retain the meanings which they have just been given, throughout the whole of § 2·7.

LEMMA 4. *If $\mathfrak{q}$ is a primary ideal of R which does not meet S, then* $\mathfrak{n} \subseteq \mathfrak{q}$.

Proof. Let $\mathfrak{q}$ be $\mathfrak{p}$-primary; then, since $\mathfrak{q}$ does not meet S, it follows that $\mathfrak{p}$ does not meet S. Let $x \in \mathfrak{n}$, then with a suitable element $c \in S$ we have $cx = 0 \in \mathfrak{q}$. But $c \notin \mathfrak{p}$, consequently $x \in \mathfrak{q}$. This completes the proof.

LEMMA 5. *No element of S^* is a zero divisor in R^*.*

Proof. If $z \in R$ we shall use z^* to denote its residue modulo $\mathfrak{n}$. Suppose that $z \in R$, that $c \in S$, and that $c^* z^* = 0$, then $cz \in \mathfrak{n}$, consequently we can find $c_1 \in S$ such that $c_1 cz = 0$. But $c_1 c \in S$, hence $z \in \mathfrak{n}$ and therefore $z^* = 0$. This completes the proof.

Now that we know that no element of S^* is a zero divisor in R^*, we can form $R^*_{S^*}$, that is, the ordinary ring of quotients of R^* with respect to S^*. We shall put $R_S = R^*_{S^*}$, and we shall call R_S either *the ring of quotients of R with respect to S*, or, if we wish to be very explicit, *the generalized ring of quotients of R with respect to S*. In the special case where S contains no zero divisors, we have $\mathfrak{n} = (0)$, which shows that the new definition of R_S agrees with the one already given in § 2·6. When we are dealing with this special situation, it is sometimes convenient to refer to R_S as *an ordinary ring of quotients*; for example, an ordinary ring of quotients of R is an extension ring of R, but a generalized ring of quotients of R need not be such.

Let us suppose that $\mathfrak{a}'$ is an ideal of R_S, then by *the contraction of $\mathfrak{a}'$ in R* we shall mean the ideal $\sigma^{-1}(\mathfrak{a}' \cap R^*)$. We shall continue to denote the contraction of $\mathfrak{a}'$ in R by $\mathfrak{a}' \cap R$, *although $\mathfrak{a}' \cap R$ is no longer an intersection*, because, as we shall see presently, this notation is both convenient and suggestive. Again, if $\mathfrak{a}$ is an ideal of R then $\mathfrak{a}^* = \sigma(\mathfrak{a})$ is an ideal of R^*, and therefore we can form the ordinary extension $R^*_{S^*}\mathfrak{a}^*$ of $\mathfrak{a}^*$ to $R^*_{S^*}$. We now put $R_S\mathfrak{a} = R^*_{S^*}\mathfrak{a}^*$ and call $R_S\mathfrak{a}$ *the extension of* $\mathfrak{a}$ *to* R_S. Since $\sigma^{-1}(\mathfrak{a}^*) = \mathfrak{a} + \mathfrak{n}$, we may, by a natural development of the notation used for residue rings, write $\mathfrak{a}^* = (\mathfrak{a} + \mathfrak{n})/\mathfrak{n}$, and then

$$R_S\mathfrak{a} = R_S[(\mathfrak{a} + \mathfrak{n})/\mathfrak{n}],$$

where the extension on the left-hand side has, as it were, a conventional meaning, but the extension on the right-hand side exists in the more ordinary sense of § 2·4. The reader should note that when R_S is an ordinary ring of quotients, the new meanings given to $\mathfrak{a}' \cap R$ and $R_S\mathfrak{a}$ agree with those which were used previously.

The main results of § 2·6, namely, Propositions 9, 10, 11 and Corollaries 1, 2 of Proposition 11, will now be shown to hold for generalized, as well as for ordinary, rings of quotients. In order to do this, it will be convenient, temporarily, to refer to the more general forms of Propositions 9, 10 and 11 as Propositions 9′, 10′ and 11′. Proposition 9′ follows at once from the definitions and Proposition 9; while Proposition 10′ follows from Propositions 7 and 10. Now let $\mathfrak{p}$ be a prime ideal of R which does not meet S and let $\mathfrak{q}$ be $\mathfrak{p}$-primary. By Lemma 4, $\mathfrak{n} \subseteq \mathfrak{q} \subseteq \mathfrak{p}$ consequently, by Proposition 5 and its corollary, $\mathfrak{p}/\mathfrak{n}$ is prime and $\mathfrak{q}/\mathfrak{n}$ is $\mathfrak{p}/\mathfrak{n}$-primary. Further, $\mathfrak{p}/\mathfrak{n}$ does not meet S^*. Proposition 11′ is now obtained by applying Proposition 11 to the ring R^* and the ideals $\mathfrak{q}/\mathfrak{n}$ and $\mathfrak{p}/\mathfrak{n}$. Again, there is a 1-1 correspondence between the prime ideals $\mathfrak{p}$ which do not meet S, and the prime ideals $\mathfrak{p}^*$ (of R^*) which do not meet S^*, such that if $\mathfrak{p}$ and $\mathfrak{p}^*$ correspond then $\mathfrak{p}^* = \mathfrak{p}/\mathfrak{n}$. If we combine this remark with Corollary 1 of Proposition 11 we at once obtain Corollary 1 of Proposition 11′. Finally, a similar argument will establish Corollary 2 of Proposition 11′.

The remainder of this section will be devoted to proving some additional properties of generalized rings of quotients. In future, whenever the reader is referred to one of Propositions 9, 10 or 11 he is to understand that we mean that Proposition in its generalized form.

PROPOSITION 12. *If* $\mathfrak{a}'_i$ $(i \in I)$ *is a family of ideals in* R_S, *then*

$$\left(\bigcap_{i\in I}\mathfrak{a}'_i\right)\cap R=\bigcap_{i\in I}\left(\mathfrak{a}'_i\cap R\right).$$

Proof
$$\left(\bigcap_{i\in I}\mathfrak{a}'_i\right)\cap R=\sigma^{-1}\left[\left(\bigcap_{i\in I}\mathfrak{a}'_i\right)\cap R^*\right]$$

$$=\sigma^{-1}\left[\bigcap_{i\in I}(\mathfrak{a}'_i\cap R^*)\right]=\bigcap_{i\in I}\sigma^{-1}(\mathfrak{a}'_i\cap R^*)=\bigcap_{i\in I}(\mathfrak{a}'_i\cap R).$$

PROPOSITION 13. *If* $\mathfrak{a}$ *and* $\mathfrak{b}$ *are ideals of* R *then*

$$R_S(\mathfrak{a}\mathfrak{b})=(R_S\mathfrak{a})(R_S\mathfrak{b}).$$

Proof. Let $\mathfrak{c}=\mathfrak{a}\mathfrak{b}$. Write $\sigma(\mathfrak{a})=\mathfrak{a}^*$, $\sigma(\mathfrak{b})=\mathfrak{b}^*$, $\sigma(\mathfrak{c})=\mathfrak{c}^*$ and $R'=R_S$, then $\mathfrak{c}^*=\mathfrak{a}^*\mathfrak{b}^*$ and therefore $R'\mathfrak{c}^*=(R'\mathfrak{a}^*)(R'\mathfrak{b}^*)$, since the extension from R^* to R' is ordinary. But $R'\mathfrak{a}^*=R_S\mathfrak{a}$, $R'\mathfrak{b}^*=R_S\mathfrak{b}$, and $R'\mathfrak{c}^*=R_S\mathfrak{c}$, consequently $R_S(\mathfrak{a}\mathfrak{b})=(R_S\mathfrak{a})(R_S\mathfrak{b})$.

PROPOSITION 14. *Let* $\mathfrak{a}$ *be an ideal of* R, *then* $(R_S\mathfrak{a})\cap R=\mathfrak{a}_S$, *where* $\mathfrak{a}_S$ *is the* S*-component*† *of* $\mathfrak{a}$.

Proof. If $z\in R$ then we shall use z^* to denote the residue of z modulo $\mathfrak{n}$. Let $z\in\mathfrak{a}_S$ and choose $c\in S$ so that $cz=a$, where $a\in\mathfrak{a}$. Then $c^*z^*=a^*$, where $a^*\in\sigma(\mathfrak{a})$ and $c^*\in S^*$, hence $z^*\in R_S\mathfrak{a}\cap R^*$, and consequently $z\in R_S\mathfrak{a}\cap R$. This shows that $\mathfrak{a}_S\subseteq R_S\mathfrak{a}\cap R$. Now let $w\in R_S\mathfrak{a}\cap R$, then $w^*\in R_S\mathfrak{a}\cap R^*$, which shows that $w^*=a_1^*/c_1^*$, where $a_1\in\mathfrak{a}$ and $c_1\in S$. Since $(wc_1)^*=a_1^*$ we have $c_1w=a_1+n$, where $n\in\mathfrak{n}$; we can therefore find $c_2\in S$ such that $c_2n=0$, and then $c_2c_1w=c_2a_1\in\mathfrak{a}$. But $c_2c_1\in S$, consequently $w\in\mathfrak{a}_S$. This shows that $R_S\mathfrak{a}\cap R\subseteq\mathfrak{a}_S$ and now the proof is complete.

PROPOSITION 15. *Suppose that*

$$\mathfrak{a}=\mathfrak{q}_1\cap\mathfrak{q}_2\cap\ldots\cap\mathfrak{q}_n,\tag{1}$$

where $\mathfrak{q}_i$ *is* $\mathfrak{p}_i$*-primary, and where* $\mathfrak{p}_i$ *meets* S *if* $m+1\leqslant i\leqslant n$ *but not if* $1\leqslant i\leqslant m$. *Put* $R'=R_S$, *and put* $\mathfrak{q}'_i=R'\mathfrak{q}_i$ *for* $1\leqslant i\leqslant m$, *then*

$$R'\mathfrak{a}=\mathfrak{q}'_1\cap\mathfrak{q}'_2\cap\ldots\cap\mathfrak{q}'_m,\tag{2}$$

and this is a primary decomposition. Further, if the decomposition (1) *is irredundant (normal) then the decomposition* (2) *is irredundant (normal).*

† See § 1·7.

Proof. By Proposition 11, if $1 \leqslant i \leqslant m$ and if $\mathfrak{p}'_i = R'\mathfrak{p}_i$, then $\mathfrak{p}'_i$ is prime and $\mathfrak{q}'_i$ is $\mathfrak{p}'_i$-primary; further, $\mathfrak{q}_i = \mathfrak{q}'_i \cap R$ and $\mathfrak{p}_i = \mathfrak{p}'_i \cap R$. Let $\mathfrak{a}' = \mathfrak{q}'_1 \cap \mathfrak{q}'_2 \cap \ldots \cap \mathfrak{q}'_m$, then this is a primary decomposition, and also, by Proposition 12,

$$\mathfrak{a}' \cap R = (\mathfrak{q}'_1 \cap R) \cap (\mathfrak{q}'_2 \cap R) \cap \ldots \cap (\mathfrak{q}'_m \cap R) = \mathfrak{q}_1 \cap \mathfrak{q}_2 \cap \ldots \cap \mathfrak{q}_m.$$

Now $$\mathfrak{q}_1 \cap \mathfrak{q}_2 \cap \ldots \cap \mathfrak{q}_m = \mathfrak{a}_S$$

(see Proposition 7 of §1·7), consequently we have $\mathfrak{a}' \cap R = \mathfrak{a}_S$. But by Proposition 14, $R_S\mathfrak{a} \cap R = \mathfrak{a}_S$, which shows that $\mathfrak{a}'$ and $R_S\mathfrak{a}$ have the same contraction in R, and from this it follows† that $\mathfrak{a}' = R_S\mathfrak{a}$, i.e. that $R'\mathfrak{a} = \mathfrak{q}'_1 \cap \mathfrak{q}'_2 \cap \ldots \cap \mathfrak{q}'_m$. Let us now suppose that the decomposition (1) is irredundant, then (2) must also be irredundant. For assume, for example, that

$$\mathfrak{q}'_1 \supseteq \mathfrak{q}'_2 \cap \mathfrak{q}'_3 \cap \ldots \cap \mathfrak{q}'_m,$$

then by contracting in R we find that

$$\mathfrak{q}_1 \supseteq \mathfrak{q}_2 \cap \mathfrak{q}_3 \cap \ldots \cap \mathfrak{q}_m \supseteq \mathfrak{q}_2 \cap \mathfrak{q}_3 \cap \ldots \cap \mathfrak{q}_n,$$

and this is a contradiction. Finally, if, in addition, $\mathfrak{p}_1, \mathfrak{p}_2, \ldots, \mathfrak{p}_n$ are all different then $\mathfrak{p}'_1, \mathfrak{p}'_2, \ldots, \mathfrak{p}'_m$ are all different, so that when the decomposition (1) is normal the decomposition (2) must be normal as well.

2·8. Local rings. Let $\mathfrak{p}$ be a proper prime ideal of a ring R and put $S = R - \mathfrak{p}$, so that S consists of all the elements of R which are not in $\mathfrak{p}$. The set S is then multiplicatively closed and it does not contain the zero element, consequently we can form the generalized ring of quotients of R with respect to S. It is customary to call the ring so obtained *the ring of quotients of R with respect to* $\mathfrak{p}$, and to denote it by $R_\mathfrak{p}$. Let us write $R' = R_\mathfrak{p}$ and $\mathfrak{p}' = R'\mathfrak{p}$, then, by Proposition 11, $\mathfrak{p}'$ is a prime ideal. Further, if $\mathfrak{p}'_1$ is any proper prime ideal of R' then, by Corollary 1 of Proposition 11, $\mathfrak{p}'_1$ is the extension of a prime ideal $\mathfrak{p}_1$ which does not meet S. We therefore have $\mathfrak{p}_1 \subseteq \mathfrak{p}$ and $\mathfrak{p}'_1 = R'\mathfrak{p}_1$, consequently $\mathfrak{p}'_1 \subseteq \mathfrak{p}'$, so that R' *has one, and only one, maximal prime ideal, and the unique maximal prime ideal is* $\mathfrak{p}'$. This is an extremely important property of a ring of quotients formed with respect to

† See Proposition 9 of §2·6.

a prime ideal, and so, in order to have our present observations expressed in a form which will be useful for future reference, we make the following definition.

DEFINITION. *A Noetherian ring which has precisely one maximal prime ideal is called a 'local ring'.*†

PROPOSITION 16. *If $\mathfrak{p}$ is a proper prime ideal of a Noetherian ring R, then the ring of quotients $R' = R_{\mathfrak{p}}$ is a local ring, and $R'\mathfrak{p}$ is its unique maximal prime ideal.*

For we have already seen that R' has only one maximal prime ideal, namely, $R'\mathfrak{p}$, and therefore it is sufficient to show that R' is Noetherian. This, however, follows from Proposition 10.

We recall that we have proved, at least for Noetherian rings, that the maximal prime ideals are the same as the maximal ideals, and we have also proved that every proper ideal is contained in at least one maximal ideal.‡ In the case of local rings these observations lead to

PROPOSITION 17. *Let R be a local ring and let $\mathfrak{p}$ be its unique maximal ideal. Then every proper ideal of R is contained in $\mathfrak{p}$. Further, the units of R are precisely the elements in $R - \mathfrak{p}$.*

Proof. The first assertion is clear from what has already been said, and the second follows from the remark that x is a unit if, and only if, the principal ideal (x) is not proper.

† The name 'Local Ring' has been given to these rings because they are used to study the local properties of irreducible varieties. The German name is *Stellenring*.

‡ See the remarks which follow Proposition 6 of §2·3.

CHAPTER III

SOME FUNDAMENTAL PROPERTIES OF NOETHERIAN RINGS

3·1. The intersection theorem. We can now describe some of the modern developments in the theory of Noetherian rings. The starting point of several of these developments has been the result which is here called the 'Intersection Theorem'. We shall therefore establish a number of different forms of this important theorem, beginning with

THEOREM 1. *Let* $\mathfrak{a}$ *be an ideal of a Noetherian ring* R, *then an element* x *of* R *belongs to* $\bigcap\limits_{i=1}^{\infty} \mathfrak{a}^i$ *if, and only if, we have* $x = ax$ *for at least one element* $a \in \mathfrak{a}$. *Further,* $\bigcap\limits_{i=1}^{\infty} \mathfrak{a}^i$ *is an isolated component of the zero ideal.*

Proof. Let S consist of all the elements which can be written in the form $1-a$ where $a \in \mathfrak{a}$, then S is multiplicatively closed. Further, if we put $\mathfrak{b} = \bigcap\limits_{i=1}^{\infty} \mathfrak{a}^i$, then the first part of the theorem asserts that $\mathfrak{b}$ is the S-component $(0)_S$ of the zero ideal; consequently the second part of the theorem follows immediately from the first. If now $x = ax$, where $a \in \mathfrak{a}$, that is, if $x \in (0)_S$, then $x = ax = a^2x = a^3x = \ldots$, which shows that x belongs to every power of $\mathfrak{a}$. We have therefore proved that $(0)_S \subseteq \mathfrak{b}$. In order to prove the opposite inclusion, we shall, as a preliminary step, show that $\mathfrak{b} \subseteq \mathfrak{a}\mathfrak{b}$. Since R is Noetherian, we may write $\mathfrak{a}\mathfrak{b} = \mathfrak{q}_1 \cap \mathfrak{q}_2 \cap \ldots \cap \mathfrak{q}_n$, where $\mathfrak{q}_i$ is $\mathfrak{p}_i$-primary. Thus for each i $\mathfrak{a}\mathfrak{b} \subseteq \mathfrak{q}_i$, and therefore either $\mathfrak{b} \subseteq \mathfrak{q}_i$ or $\mathfrak{a} \subseteq \mathfrak{p}_i$. But if $\mathfrak{a} \subseteq \mathfrak{p}_i$ we can, by the Corollary of Proposition 8 of § 1·9, find an integer n such that $\mathfrak{p}_i^n \subseteq \mathfrak{q}_i$, and then $\mathfrak{b} \subseteq \mathfrak{a}^n \subseteq \mathfrak{p}_i^n \subseteq \mathfrak{q}_i$, so that *in any case* $\mathfrak{b} \subseteq \mathfrak{q}_i$. Since, for each i, $\mathfrak{b}$ is contained in $\mathfrak{q}_i$ we have $\mathfrak{b} \subseteq \mathfrak{a}\mathfrak{b}$, and to complete the proof we have only to deduce that $\mathfrak{b} \subseteq (0)_S$. The proposition which follows shows that this deduction can be made.

PROPOSITION 1. *Suppose that* $\mathfrak{a}$ *and* $\mathfrak{b}$ *are two ideals in a*

Noetherian ring R, and suppose, also, that $\mathfrak{b} \subseteq \mathfrak{a}\mathfrak{b}$. Then there exists an element $a \in \mathfrak{a}$ such that $(1-a)\mathfrak{b} = (0)$.

Proof. Since R is Noetherian $\mathfrak{b}$ is finitely generated, say $\mathfrak{b} = (b_1, b_2, \ldots, b_n)$. Then $b_i \in \mathfrak{a}\mathfrak{b}$, hence

$$b_i = a_{i1} b_1 + a_{i2} b_2 + \ldots + a_{in} b_n,$$

where the a_{ij} belong to $\mathfrak{a}$, consequently $\sum_{r=1}^{n} (\delta_{ir} - a_{ir}) b_r = 0$, where $\delta_{ir} = 1$ if $i = r$, and $\delta_{ir} = 0$ if $i \neq r$. If, with the usual notation for determinants, $\Delta = | \delta_{ij} - a_{ij} |$, then $\Delta b_r = 0$ for each r, and therefore $\Delta \mathfrak{b} = (0)$. But $\Delta = 1 - a$, where $a \in \mathfrak{a}$, and this completes the proof.

Let us note that $(0)_S = (0)$ if, and only if, S contains no zero divisor. This gives us the following special case of Theorem 1.

THEOREM 2. *If $\mathfrak{a}$ is an ideal of a Noetherian ring R then $\bigcap_{i=1}^{\infty} \mathfrak{a}^i = (0)$ if, and only if, there is no zero divisor of R which is congruent, modulo $\mathfrak{a}$, to the unit element.*

An immediate consequence of Theorem 2 is

THEOREM 3. *Let R be a local ring and let $\mathfrak{p}$ be its unique maximal ideal, then $\bigcap_{i=1}^{\infty} \mathfrak{p}^i = (0)$.*

Proof. Suppose that $x \equiv 1 \pmod{\mathfrak{p}}$ then $x \notin \mathfrak{p}$, and therefore (by Proposition 17 of §2·8) x is a unit. This shows, in particular, that x is not a zero divisor, consequently there is no zero divisor which is congruent, modulo $\mathfrak{p}$, to the unit element. Theorem 3 now follows from Theorem 2.

3·2. Symbolic prime powers. Let $\mathfrak{p}$ be a proper prime ideal in a Noetherian ring R and let n be a positive integer. We know† that $\mathfrak{p}^n$ need not be $\mathfrak{p}$-primary, but we can say, at least, that $\mathfrak{p}$ is a minimal prime ideal of $\mathfrak{p}^n$, because every prime ideal which contains $\mathfrak{p}^n$ must necessarily contain $\mathfrak{p}$. This shows‡ that the $\mathfrak{p}$-primary component of $\mathfrak{p}^n$ is the same in all normal decompositions of $\mathfrak{p}^n$. We shall denote this uniquely determined

† See Example 3 at the end of Chapter I.
‡ See the Corollary of Theorem 3 in §1·7.

$\mathfrak{p}$-primary component by $\mathfrak{p}^{(n)}$, and we shall call it *the nth symbolic prime power of* $\mathfrak{p}$. Before we proceed further, let us note that if $S = R - \mathfrak{p}$ then, by Proposition 7 of § 1·7, $\mathfrak{p}^{(n)}$ is the S-component of $\mathfrak{p}^n$. Let $R' = R_S$ be the generalized ring of quotients of R with respect to S, and put $\mathfrak{p}' = R'\mathfrak{p}$, so that R' is a local ring with $\mathfrak{p}'$ as its maximal ideal. By Proposition 13 of § 2·7, $R'\mathfrak{p}^r = \mathfrak{p}'^r$, so that, by Proposition 14 of § 2·7, $\mathfrak{p}'^r \cap R$ is the S-component of $\mathfrak{p}^r$; that is, $\mathfrak{p}'^r \cap R = \mathfrak{p}^{(r)}$. It now follows, from Proposition 12 of § 2·7, that

$$\bigcap_{r=1}^{\infty} \mathfrak{p}^{(r)} = \bigcap_{r=1}^{\infty} (\mathfrak{p}'^r \cap R) = \left(\bigcap_{r=1}^{\infty} \mathfrak{p}'^r\right) \cap R,$$

and, by Theorem 3, $\bigcap_{r=1}^{\infty} \mathfrak{p}'^r$ is the zero ideal of R'. Thus we have proved that the intersection of all the symbolic prime powers is just the contraction in R of the zero ideal of R'. But, by the definitions, this contraction is none other than the S-component of the zero ideal of R, consequently we have established the following theorem:

THEOREM 4. *Let $\mathfrak{p}$ be a proper prime ideal in a Noetherian ring R, and let $\mathfrak{p}^{(r)}$ be its rth symbolic prime power. Then $\bigcap_{r=1}^{\infty} \mathfrak{p}^{(r)}$ is that component of the zero ideal which is determined by the multiplicatively closed set $R - \mathfrak{p}$.*

3·3. The length of a primary ideal. Suppose that $\mathfrak{q}$ is a $\mathfrak{p}$-primary ideal, and suppose also that $\mathfrak{q}_1, \mathfrak{q}_2, \ldots, \mathfrak{q}_n$ are $\mathfrak{p}$-primary ideals such that $\mathfrak{p} = \mathfrak{q}_1 \supset \mathfrak{q}_2 \supset \ldots \supset \mathfrak{q}_n = \mathfrak{q}$, where all the inclusions are strict. We shall describe this situation by saying that $[\mathfrak{q}_1, \mathfrak{q}_2, \ldots, \mathfrak{q}_n]$ is a primary chain from $\mathfrak{p}$ to $\mathfrak{q}$. If now $[\mathfrak{q}_1, \mathfrak{q}_2, \ldots, \mathfrak{q}_n]$ and $[\mathfrak{q}'_1, \mathfrak{q}'_2, \ldots, \mathfrak{q}'_m]$ are two primary chains from $\mathfrak{p}$ to $\mathfrak{q}$, then we shall say that $[\mathfrak{q}'_1, \mathfrak{q}'_2, \ldots, \mathfrak{q}'_m]$ is a *refinement* of $[\mathfrak{q}_1, \mathfrak{q}_2, \ldots, \mathfrak{q}_n]$ if every $\mathfrak{q}_i$ occurs among the $\mathfrak{q}'_j$; moreover, if, in addition, the two chains are not identical then we shall say that $[\mathfrak{q}'_1, \mathfrak{q}'_2, \ldots, \mathfrak{q}'_m]$ is a *proper refinement* of $[\mathfrak{q}_1, \mathfrak{q}_2, \ldots, \mathfrak{q}_n]$. A primary chain from $\mathfrak{p}$ to $\mathfrak{q}$ will be called a *composition series* for $\mathfrak{q}$ if it has no proper refinements. If $\mathfrak{q} = \mathfrak{p}$ then there is only one primary chain from $\mathfrak{p}$ to $\mathfrak{q}$, namely, $[\mathfrak{p}]$, and this is also a composition series for $\mathfrak{p}$. The present section and the one which follows will be devoted to proving the following theorem:

THEOREM 5. *Let $\mathfrak{p}$ be a proper prime ideal in a Noetherian ring R and let $\mathfrak{q}$ be $\mathfrak{p}$-primary. Then there exists at least one composition series for $\mathfrak{q}$, and any two such composition series contain the same number of terms. Further, every primary chain from $\mathfrak{p}$ to $\mathfrak{q}$ can be refined into a composition series for $\mathfrak{q}$.*

Once this theorem has been proved we shall be in a position to make the following

DEFINITION. *If $\mathfrak{q}$ is a primary ideal in a Noetherian ring, then the number of terms in a composition series for $\mathfrak{q}$ will be called the 'length' of $\mathfrak{q}$.*

For example, the length of a prime ideal is always one. This concept of the length of a primary ideal has close connexions with the concept of an intersection multiplicity in algebraic geometry.

As the proof of Theorem 5 will not be given immediately, but only after a certain amount of preliminary investigation has been completed, we shall explain here what are the main stages of the discussion. We shall reduce Theorem 5 to a special case by showing that we may suppose that $\mathfrak{q}$ is the zero ideal in what is known as a *primary ring*. This reduction will be made in the present section, and then in the section which follows we shall prove the special case of Theorem 5 to which we have referred. There is, however, a general result which is used to investigate the structure of primary rings, and which, because of its generality, it is more convenient to prove at once, before we begin the study of very specialized situations. This result is

PROPOSITION 2. *Let $\mathfrak{a}$ be an ideal in a Noetherian ring R and let $\mathfrak{p}$ be any maximal prime ideal of R. Then $\mathfrak{a}/\mathfrak{a}\mathfrak{p}$ forms, in a natural way, a finite-dimensional vector space over the field $R/\mathfrak{p}$. Further, if $\mathfrak{b}$ is an ideal satisfying $\mathfrak{a} \supseteq \mathfrak{b} \supseteq \mathfrak{a}\mathfrak{p}$, then $\mathfrak{b}/\mathfrak{a}\mathfrak{p}$ is a vector subspace of $\mathfrak{a}/\mathfrak{a}\mathfrak{p}$.*

Proof. By $\mathfrak{a}/\mathfrak{a}\mathfrak{p}$ we mean, of course, all the residue classes of $\mathfrak{a}\mathfrak{p}$ which are determined by elements of $\mathfrak{a}$, or, in other words, we mean the residue classes of $\mathfrak{a}\mathfrak{p}$ which are composed entirely of elements of $\mathfrak{a}$. Suppose that $u \in \mathfrak{a}/\mathfrak{a}\mathfrak{p}$ and that $a \in \mathfrak{a}$ is a representative of u; suppose, also, that $\rho \in R/\mathfrak{p}$ and that r is a representative of ρ. Then the residue class of ra modulo $\mathfrak{a}\mathfrak{p}$ depends only on ρ and u, consequently we may, without ambiguity, denote this residue class by ρu. To see the truth of this assertion, let us suppose that a' and r' are alternative representatives of u and ρ respectively, so that $a \equiv a' \pmod{\mathfrak{a}\mathfrak{p}}$ and $r \equiv r' \pmod{\mathfrak{p}}$. Then $ra - r'a' = a(r-r') + r'(a-a') \equiv 0 \pmod{\mathfrak{a}\mathfrak{p}}$, and therefore ra and $r'a'$ determine the same residue class modulo $\mathfrak{a}\mathfrak{p}$. Now let $\rho, \rho' \in R/\mathfrak{p}$ and let $u, u' \in \mathfrak{a}/\mathfrak{a}\mathfrak{p}$. It is clear that $\rho u \in \mathfrak{a}/\mathfrak{a}\mathfrak{p}$, and it is easily verified that $\rho(u+u') = \rho u + \rho u'$, that $(\rho+\rho')u = \rho u + \rho' u$, and that $\rho(\rho' u) = (\rho\rho')u$. Moreover, since the unit element of R is a representative of the unit element of $R/\mathfrak{p}$, it follows that if we multiply u by the unit element of $R/\mathfrak{p}$ then we shall obtain just u itself. Thus, with the definition of the product ρu which was given above, $\mathfrak{a}/\mathfrak{a}\mathfrak{p}$ is a vector space over $R/\mathfrak{p}$. Let $\mathfrak{a} = (a_1, a_2, \dots, a_m)$, and let u_i be the residue of a_i modulo $\mathfrak{a}\mathfrak{p}$. Since every element $a \in \mathfrak{a}$ can be written in the form $a = r_1 a_1 + r_2 a_2 + \dots + r_m a_m$, where $r_i \in R$, it follows that every element $u \in \mathfrak{a}/\mathfrak{a}\mathfrak{p}$ can be written in the form $u = \alpha_1 u_1 + \alpha_2 u_2 + \dots + \alpha_m u_m$, where $\alpha_i \in R/\mathfrak{p}$. This shows that as a vector space $\mathfrak{a}/\mathfrak{a}\mathfrak{p}$ is of finite dimension. Finally, assume that the ideal $\mathfrak{b}$ satisfies $\mathfrak{a} \supseteq \mathfrak{b} \supseteq \mathfrak{a}\mathfrak{p}$, let $v_1, v_2 \in \mathfrak{b}/\mathfrak{a}\mathfrak{p}$, and let $\alpha \in R/\mathfrak{p}$. Then $v_1 + v_2 \in \mathfrak{b}/\mathfrak{a}\mathfrak{p}$ and also $\alpha v_1 \in \mathfrak{b}/\mathfrak{a}\mathfrak{p}$, consequently $\mathfrak{b}/\mathfrak{a}\mathfrak{p}$ is a vector subspace of $\mathfrak{a}/\mathfrak{a}\mathfrak{p}$.

After this digression, we return to the problem of reducing Theorem 5 to a comparatively simple special case. To do this let us suppose that the hypotheses of Theorem 5 are all satisfied, let us put $S = R - \mathfrak{p}$, and let us denote by R' the generalized ring of quotients of R with respect to S. If we now put $\mathfrak{p}' = R'\mathfrak{p}$ and $\mathfrak{q}' = R'\mathfrak{q}$, then, by Proposition 11 of §2·6, $\mathfrak{p}'$ is prime and $\mathfrak{q}'$ is $\mathfrak{p}'$-primary. Moreover, by Corollary 2 of the same proposition, there is a 1-1 correspondence between the primary chains $[\mathfrak{q}_1, \mathfrak{q}_2, \dots, \mathfrak{q}_m]$ from $\mathfrak{p}$ to $\mathfrak{q}$ and the primary chains $[\mathfrak{q}'_1, \mathfrak{q}'_2, \dots, \mathfrak{q}'_n]$ from $\mathfrak{p}'$ to $\mathfrak{q}'$. Also, the correspondence is such that if $[\mathfrak{q}_1, \mathfrak{q}_2, \dots, \mathfrak{q}_m]$ and $[\mathfrak{q}'_1, \mathfrak{q}'_2, \dots, \mathfrak{q}'_n]$ correspond then $m = n$, and for each i we have $\mathfrak{q}'_i \cap R = \mathfrak{q}_i$ and $R'\mathfrak{q}_i = \mathfrak{q}'_i$. This shows that it is sufficient to prove Theorem 5 when the ring is R', when the prime ideal is $\mathfrak{p}'$, and when the primary ideal is $\mathfrak{q}'$. Now, by Proposition 16 of §2·8, R' is a local ring and $\mathfrak{p}'$ is its maximal ideal. Thus, *in proving Theorem* 5, *we may, without loss of generality, assume that R is a local ring, and that $\mathfrak{p}$ is its maximal ideal.* Let us make these assumptions, and let us suppose that $\mathfrak{a}$ is an ideal between $\mathfrak{q}$ and $\mathfrak{p}$ so that $\mathfrak{q} \subseteq \mathfrak{a} \subseteq \mathfrak{p}$.

Since $\mathfrak{q}$ is $\mathfrak{p}$-primary we can find an integer r such that $\mathfrak{p}^r \subseteq \mathfrak{q} \subseteq \mathfrak{a} \subseteq \mathfrak{p}$; but $\mathfrak{p}$ is a maximal ideal and therefore, by Proposition 9 of § 1·9, $\mathfrak{a}$ must be $\mathfrak{p}$-primary. In different language, this means that the $\mathfrak{p}$-primary ideals between $\mathfrak{q}$ and $\mathfrak{p}$ are simply the ideals between $\mathfrak{q}$ and $\mathfrak{p}$.

We can now make a further simplification. Since every proper prime ideal which contains $\mathfrak{q}$ must contain $\mathfrak{p}$, and since $\mathfrak{p}$ is maximal, it follows that $\mathfrak{p}$ is the *only* proper prime ideal which contains $\mathfrak{q}$. This shows that $R/\mathfrak{q}$ is a Noetherian ring which has only one proper prime ideal, namely, $\mathfrak{p}/\mathfrak{q}$. Also there is a 1-1 correspondence between chains of ideals from $\mathfrak{p}$ to $\mathfrak{q}$, and chains of ideals from $\mathfrak{p}/\mathfrak{q}$ to the zero ideal of $R/\mathfrak{q}$. In order to take advantage of this situation, we make the following

DEFINITION. *A Noetherian ring, which has only one proper prime ideal, will be called a 'primary ring'.*

3·4. Primary rings. Throughout the whole of this section R will denote a primary ring, and $\mathfrak{p}$ will denote its unique proper prime ideal. Since there is no proper prime ideal besides $\mathfrak{p}$, it follows that every proper ideal is $\mathfrak{p}$-primary. It also follows that $\mathfrak{p}$ is a maximal prime ideal and that $R/\mathfrak{p}$ is a field.

Suppose that $\mathfrak{a}$ and $\mathfrak{b}$ are proper ideals of R and that $\mathfrak{a} \supseteq \mathfrak{b}$. A set $\mathfrak{c}_1, \mathfrak{c}_2, \ldots, \mathfrak{c}_n$ of ideals such that $\mathfrak{a} = \mathfrak{c}_1 \supset \mathfrak{c}_2 \supset \ldots \supset \mathfrak{c}_n = \mathfrak{b}$, where the inclusions are strict, will be called *a chain from* $\mathfrak{a}$ *to* $\mathfrak{b}$. We define refinements and proper refinements of such chains in the obvious way, and a chain from $\mathfrak{a}$ to $\mathfrak{b}$ which has no proper refinements, will be called *a composition series from* $\mathfrak{a}$ *to* $\mathfrak{b}$. If we speak of a composition series without any mention at all of its beginning or of its end, we shall always mean a composition series from $\mathfrak{p}$ to (0). The remarks of § 3·3 show that, in order to prove Theorem 5, it is sufficient to prove the following result concerning primary rings:

THEOREM A. *There exists at least one composition series, and any two composition series contain the same number of terms. Further, every chain from* $\mathfrak{p}$ *to* (0) *can be refined into a composition series.*

The proof of Theorem A is divided into two lemmas.

LEMMA 1. *There is at least one composition series from* $\mathfrak{p}$ *to* (0).

Proof. We shall show, first, that for each positive integer n there is a composition series from $\mathfrak{p}^n$ to $\mathfrak{p}^{n+1}$. Suppose that $\mathfrak{p}^n = \mathfrak{a}_1 \supset \mathfrak{a}_2 \supset \ldots \supset \mathfrak{a}_r = \mathfrak{p}^{n+1}$ is a chain from $\mathfrak{p}^n$ to $\mathfrak{p}^{n+1}$. By Proposition 2, $\mathfrak{p}^n/\mathfrak{p}^{n+1}$ is a vector space over $R/\mathfrak{p}$ of dimension l_n (say), and, by the same proposition, $\mathfrak{a}_\nu/\mathfrak{p}^{n+1}$ is a subspace of dimension d_ν (say). We therefore have $l_n = d_1 > d_2 > \ldots > d_r \geqslant 0$, consequently r cannot exceed $l_n + 1$. This shows that, for a fixed n, we can construct a *longest possible* chain from $\mathfrak{p}^n$ to $\mathfrak{p}^{n+1}$. Such a chain cannot have any proper refinements, and therefore it must be a composition series from $\mathfrak{p}^n$ to $\mathfrak{p}^{n+1}$. To complete the proof of the lemma, we note that as the ideal (0) is proper it must be $\mathfrak{p}$-primary, consequently we can find an integer l such that $\mathfrak{p}^l = (0)$. If now for each n which satisfies $1 \leqslant n < l$ we construct a composition series from $\mathfrak{p}^n$ to $\mathfrak{p}^{n+1}$, then by combining these composition series we shall obtain a composition series from $\mathfrak{p}$ to (0).

LEMMA 2. *Let* $\mathfrak{p} = \mathfrak{a}_r \supset \mathfrak{a}_{r-1} \supset \ldots \supset \mathfrak{a}_0 = (0)$ *be a chain from* $\mathfrak{p}$ *to* (0), *and let* $\mathfrak{p} = \mathfrak{b}_l \supset \mathfrak{b}_{l-1} \supset \ldots \supset \mathfrak{b}_0 = (0)$ *be a composition series, then* $r \leqslant l$.

Proof. We may suppose that $l \geqslant 1$, for if $l = 0$ then $\mathfrak{p} = (0)$ and hence $r = 0$. Assuming then that $l \geqslant 1$, we note that $\mathfrak{b}_1 \nsubseteq \mathfrak{a}_0$ and that $\mathfrak{b}_1 \subseteq \mathfrak{a}_r$, consequently there exists an integer k such that $\mathfrak{b}_1 \nsubseteq \mathfrak{a}_k$ and $\mathfrak{b}_1 \subseteq \mathfrak{a}_{k+1}$. We shall now show that

$$\mathfrak{b}_1 + \mathfrak{a}_k \supset \mathfrak{b}_1 + \mathfrak{a}_{k-1} \supset \ldots \supset \mathfrak{b}_1 + \mathfrak{a}_0 = \mathfrak{b}_1. \tag{1}$$

To do this we suppose that $0 \leqslant j < k$, and we choose $a_{j+1} \in \mathfrak{a}_{j+1}$ so that $a_{j+1} \notin \mathfrak{a}_j$. Then $a_{j+1} \notin \mathfrak{a}_j + \mathfrak{b}_1$, for if the contrary were true we could find $a_j \in \mathfrak{a}_j$ such that $a_{j+1} - a_j \in \mathfrak{b}_1 \cap \mathfrak{a}_{j+1}$. But, since $\mathfrak{b}_1 \nsubseteq \mathfrak{a}_{j+1}$, $\mathfrak{b}_1 \cap \mathfrak{a}_{j+1}$ is strictly contained in $\mathfrak{b}_1$ and therefore $\mathfrak{b}_1 \cap \mathfrak{a}_{j+1} = (0)$. Thus $a_{j+1} = a_j \in \mathfrak{a}_j$ and we have a contradiction. From $a_{j+1} \notin \mathfrak{a}_j + \mathfrak{b}_1$ it follows that $\mathfrak{a}_{j+1} + \mathfrak{b}_1 \supset \mathfrak{a}_j + \mathfrak{b}_1$ and this proves (1). We next assert that *there exists a chain from* $\mathfrak{p}$ *to* (0) *which has at least* $r+1$ *terms, and which ends with* $\mathfrak{b}_1$, (0); in fact, to obtain such a chain we have only to strike out the repetitions in

$$\mathfrak{p} = \mathfrak{a}_r \supset \mathfrak{a}_{r-1} \supset \ldots \supset \mathfrak{a}_{k+1} \supseteq \mathfrak{a}_k + \mathfrak{b}_1 \supset \ldots \supset \mathfrak{a}_0 + \mathfrak{b}_1 = \mathfrak{b}_1 \supset (0).$$

Now in $R/\mathfrak{b}_1$, $\mathfrak{b}_l/\mathfrak{b}_1 \supset \mathfrak{b}_{l-1}/\mathfrak{b}_1 \supset \ldots \supset \mathfrak{b}_1/\mathfrak{b}_1$ is a composition series, and, by the above remarks, there exists a chain from $\mathfrak{p}/\mathfrak{b}_1$ to the zero ideal which has at least r terms. But $R/\mathfrak{b}_1$ is a primary ring, consequently, if $l-1 \geqslant 1$, we can repeat the above argument. If we do this and then return to the original ring R we obtain the following: *there exists a chain from* $\mathfrak{p}$ *to* (0) *which has at least* $r+1$ *terms, and which ends with* $\mathfrak{b}_2, \mathfrak{b}_1, (0)$. We next consider the ring $R/\mathfrak{b}_2$, and it is now clear that, continuing in this way, we shall arrive at a chain from $\mathfrak{p}$ to (0) which has at least $r+1$ terms, and which ends with $\mathfrak{b}_l, \mathfrak{b}_{l-1}, \ldots, \mathfrak{b}_1, (0)$. But such a chain can only be $\mathfrak{b}_l \supset \mathfrak{b}_{l-1} \supset \ldots \supset \mathfrak{b}_0$ itself, and this shows that $r \leqslant l$.

COROLLARY 1. *Any two composition series have the same number of terms.*

For by the lemma neither composition series can have more terms than the other.

COROLLARY 2. *Any chain* $\mathfrak{p} = \mathfrak{a}_1 \supset \mathfrak{a}_2 \supset \ldots \supset \mathfrak{a}_m = (0)$ *can be refined into a composition series.*

Proof. Let k be the number of terms in a composition series, then, by the lemma, no refinement of $\mathfrak{p} = \mathfrak{a}_1 \supset \mathfrak{a}_2 \supset \ldots \supset \mathfrak{a}_m = (0)$ can have more than k terms; consequently we can find a *longest possible* refinement. Such a refinement will be a composition series.

Theorem A, and with it Theorem 5, now follows by combining Lemma 1 with the corollaries of Lemma 2.

3·5. Rank and dimension. Suppose that $\mathfrak{p}$ is a prime ideal in the polynomial ring $C[x_1, x_2, \ldots, x_n]$, where C is the field of complex numbers, then a well-known theorem in algebraic geometry enables us to associate with $\mathfrak{p}$ a certain algebraic locus V in n-dimensional space. This locus V will be irreducible, and since irreducible loci can be classed as curves, as surfaces, and so on, V will have a certain dimension. If V is of dimension d then we say that $\mathfrak{p}$ is a prime ideal of dimension d, and we call the complementary dimension, that is, $n-d$, the rank of $\mathfrak{p}$. In this section we shall be concerned with the concepts of rank and dimension in Noetherian rings, but, before we commence the strict theory, it is advisable to notice some respects in which the geometric situation differs from the more general situation with which we shall be dealing. For any prime ideal in $C[x_1, x_2, \ldots, x_n]$

the sum of the rank and the dimension is always the same (namely, n) and therefore if we know either the rank or the dimension of a prime ideal then we know both these numbers. In the general case, however, we have no such simple situation, and we may say that the concepts of rank and dimension have a certain kind of independence. In the problems which we shall consider, it will usually be the rank which is the more important. We come now to the actual definitions.

DEFINITION. *A proper prime ideal $\mathfrak{p}$ is said to be of 'rank r' if there exists a descending chain of r different prime ideals all of which are strictly contained in $\mathfrak{p}$; and if, at the same time, there is no such chain with $r+1$ members.*

DEFINITION. *A proper prime ideal $\mathfrak{p}$ is said to be of 'dimension d' if there exists an ascending chain of d different proper prime ideals all of which strictly contain $\mathfrak{p}$; and if, at the same time, there is no such chain with $d+1$ members.*

Suppose that $\mathfrak{p}$ and $\mathfrak{p}'$ are proper prime ideals and that $\mathfrak{p} \subset \mathfrak{p}'$. It follows at once from the definitions that if the rank of $\mathfrak{p}$ is finite then the rank of $\mathfrak{p}'$ is greater than the rank of $\mathfrak{p}$; it also follows that if the dimension of $\mathfrak{p}'$ is finite then the dimension of $\mathfrak{p}$ is greater than the dimension of $\mathfrak{p}'$. Another easily obtained consequence of the definitions, which we record for future reference, is

PROPOSITION 3. *Let S be a multiplicatively closed set (not containing the zero element) in an arbitrary ring R and let $\mathfrak{p}$ be a prime ideal not meeting S, then $\mathfrak{p}$ and its extension in the ring R_S of quotients have the same rank.*

This follows at once from Corollary 1 of Proposition 11 of §2·6.

In the case of a Noetherian ring we extend our definitions of rank and dimension in order to make them applicable to an arbitrary proper ideal.

DEFINITION. *Let $\mathfrak{a}$ be a proper ideal in a Noetherian ring and let $\mathfrak{p}_1, \mathfrak{p}_2, \ldots, \mathfrak{p}_t$ be the prime ideals which belong to $\mathfrak{a}$. Then by the rank of $\mathfrak{a}$ we mean the smallest of the ranks of the prime ideals $\mathfrak{p}_1, \mathfrak{p}_2, \ldots, \mathfrak{p}_t$, and by the dimension of $\mathfrak{a}$ we mean the largest of the dimensions of $\mathfrak{p}_1, \mathfrak{p}_2, \ldots, \mathfrak{p}_t$.*

It is clear that in determining the rank and the dimension of an ideal $\mathfrak{a}$, only the *minimal* prime ideals of $\mathfrak{a}$ need be taken into account. Suppose now that $\mathfrak{a}$ and $\mathfrak{a}'$ are two proper ideals in a Noetherian ring, suppose also that $\mathfrak{a} \subseteq \mathfrak{a}'$, and let $\mathfrak{p}'$ be a minimal prime ideal of $\mathfrak{a}'$. Then $\mathfrak{p}' \supseteq \mathfrak{a}' \supseteq \mathfrak{a}$ and therefore, by Theorem 1 of § 1·6, $\mathfrak{p}'$ contains a minimal prime ideal $\mathfrak{p}$ of $\mathfrak{a}$. From this and from our earlier remarks, it follows that the dimension of $\mathfrak{a}$ is at least as great as the dimension of $\mathfrak{a}'$ and it also follows that the rank of $\mathfrak{a}'$ is at least as great as the rank of $\mathfrak{a}$. All the deeper results, concerning rank and dimension, which we shall prove will depend on the following theorem:

THEOREM 6. *Let b be a non-unit in a Noetherian ring R and let $\mathfrak{p}$ be a minimal prime ideal of the principal ideal (b), then the rank of $\mathfrak{p}$ is at most one.*

Proof. We shall suppose that $\mathfrak{p} \supset \mathfrak{p}_1 \supseteq \mathfrak{p}_2$, where $\mathfrak{p}_1$ and $\mathfrak{p}_2$ are prime, and we shall deduce from this that $\mathfrak{p}_1 = \mathfrak{p}_2$. By passing to the ring of quotients of R with respect to $R - \mathfrak{p}$ the problem becomes greatly simplified, the simplification amounting, in fact, to this: *For the purposes of the proof, we may assume that (b) is $\mathfrak{p}$-primary, and we may also assume that R is a local ring which has $\mathfrak{p}$ as its maximal ideal.* From now on all these assumptions will be made. Let $\mathfrak{p}_1^{(r)}$ be the rth symbolic prime power of $\mathfrak{p}_1$. Since (b) is $\mathfrak{p}$-primary and since $\mathfrak{p}$ is maximal, it follows† that any ideal between $\mathfrak{p}$ and (b) is $\mathfrak{p}$-primary, and, in particular, it follows that $(b) + \mathfrak{p}_1^{(r)}$ is $\mathfrak{p}$-primary. By Theorem 5, the number of different terms in the descending chain

$$\mathfrak{p}_1^{(1)} + (b) \supseteq \mathfrak{p}_1^{(2)} + (b) \supseteq \mathfrak{p}_1^{(3)} + (b) \supseteq \dots$$

is finite, because it cannot exceed the length of the primary ideal (b), consequently we have

$$\mathfrak{p}_1^{(s)} + (b) = \mathfrak{p}_1^{(s+1)} + (b) = \mathfrak{p}^{(s+2)} + (b) = \dots$$

for a suitable integer s. In what follows m denotes an integer which is not less than s. We assert that

$$\mathfrak{p}_1^{(m)} \subseteq (b)\, \mathfrak{p}_1^{(m)} + \mathfrak{p}_1^{(m+1)}. \tag{1}$$

† See Proposition 9 of § 1·9.

For let $x \in \mathfrak{p}_1^{(m)}$ then $x \in (b) + \mathfrak{p}_1^{(m)} = (b) + \mathfrak{p}_1^{(m+1)}$, and therefore $x = rb + y$, where $r \in R$ and where $y \in \mathfrak{p}_1^{(m+1)}$. We then have $br = x - y \in \mathfrak{p}_1^{(m)}$, consequently, since $b \notin \mathfrak{p}_1$ and since $\mathfrak{p}_1^{(m)}$ is $\mathfrak{p}_1$-primary, $r \in \mathfrak{p}_1^{(m)}$. This shows that $x = br + y \in (b)\,\mathfrak{p}_1^{(m)} + \mathfrak{p}_1^{(m+1)}$ and establishes (1). We shall now show that

$$\mathfrak{p}_1^{(m)} = \mathfrak{p}_1^{(m+1)}, \tag{2}$$

and for this it will be sufficient to show that $\mathfrak{p}_1^{(m)} \subseteq \mathfrak{p}_1^{(m+1)}$. Let $\mathfrak{p}_1^{(m)} = (u_1, u_2, \ldots, u_k)$, then, by (1),

$$u_i \equiv b(r_{i1}u_1 + r_{i2}u_2 + \ldots + r_{ik}u_k) \quad (\operatorname{mod} \mathfrak{p}_1^{(m+1)}),$$

where the r_{ij} are suitable elements of R. Let δ_{ij} have its usual meaning, then

$$\sum_{j=1}^{k} (\delta_{ij} - br_{ij})\, u_j \equiv 0 \quad (\operatorname{mod} \mathfrak{p}_1^{(m+1)}),$$

and therefore for each j we have $\Delta u_j \equiv 0 \, (\operatorname{mod} \mathfrak{p}_1^{(m+1)})$, where Δ is the determinant $|\, \delta_{ij} - br_{ij} \,|$. Thus

$$\Delta \mathfrak{p}_1^{(m)} \subseteq \mathfrak{p}_1^{(m+1)}.$$

Now
$$\Delta \equiv 1 \quad (\operatorname{mod} (b)),$$

so *a fortiori* $\Delta \equiv 1 \, (\operatorname{mod} \mathfrak{p})$. This shows that Δ is a unit, and therefore, since $\Delta \mathfrak{p}_1^{(m)} \subseteq \mathfrak{p}_1^{(m+1)}$, we have $\mathfrak{p}_1^{(m)} \subseteq \mathfrak{p}_1^{(m+1)}$, so that (2) is now established. But (2) holds for all $m \geqslant s$, consequently

$$\mathfrak{p}_1^s \subseteq \mathfrak{p}_1^{(s)} = \mathfrak{p}_1^{(s+1)} = \mathfrak{p}_1^{(s+2)} = \ldots$$

so that, by Theorem 4, $\mathfrak{p}_1^s \subseteq \mathfrak{n}$, *where* $\mathfrak{n}$ *is the component of the zero ideal determined by* $R - \mathfrak{p}_1$. We recall that we wish to prove that $\mathfrak{p}_1 = \mathfrak{p}_2$. Let $x \in \mathfrak{p}_1$ then $x^s \in \mathfrak{p}_1^s \subseteq \mathfrak{n}$, and therefore we can find $c \notin \mathfrak{p}_1$ such that $cx^s = 0$. Since $c \notin \mathfrak{p}_1$, we have *a fortiori* $c \notin \mathfrak{p}_2$ consequently from $cx^s = 0 \in \mathfrak{p}_2$ it follows that $x \in \mathfrak{p}_2$. Thus $\mathfrak{p}_1 \subseteq \mathfrak{p}_2$ and therefore $\mathfrak{p}_1 = \mathfrak{p}_2$. The proof is now complete.

PROPOSITION 4. *Let R be a Noetherian ring, and let $\mathfrak{p}_0 \subset \mathfrak{p}$ be proper prime ideals such that there exists at least one prime ideal strictly between $\mathfrak{p}_0$ and $\mathfrak{p}$. Then, given any finite set $\mathfrak{p}'_1, \mathfrak{p}'_2, \ldots, \mathfrak{p}'_k$ of prime ideals, none of which contains $\mathfrak{p}$, there exists a prime ideal $\mathfrak{p}^*$, which is not contained in any $\mathfrak{p}'_i$, and which satisfies $\mathfrak{p}_0 \subset \mathfrak{p}^* \subset \mathfrak{p}$.*

Proof. Suppose that $\mathfrak{p}_0 \subset \mathfrak{p}_1 \subset \mathfrak{p}$, where $\mathfrak{p}_1$ is prime. The ideal $\mathfrak{p}$ is not contained in any of $\mathfrak{p}_0, \mathfrak{p}'_1, \ldots, \mathfrak{p}'_k$, consequently, by Proposition 6 of § 1·5, we can find $a \in \mathfrak{p}$ such that a does not belong to any of $\mathfrak{p}_0, \mathfrak{p}'_1, \ldots, \mathfrak{p}'_k$. Since $(a) + \mathfrak{p}_0 \subseteq \mathfrak{p}$, we can find a minimal prime ideal $\mathfrak{p}^*$ of $(a) + \mathfrak{p}_0$ such that $\mathfrak{p}^* \subseteq \mathfrak{p}$. By construction, $\mathfrak{p}^* \supset \mathfrak{p}_0$ and $\mathfrak{p}^*$ is not contained in any $\mathfrak{p}'_i$, hence the proposition will be proved if we show that $\mathfrak{p}^* \neq \mathfrak{p}$. We shall assume that $\mathfrak{p}^* = \mathfrak{p}$, and from this we shall obtain a contradiction. Since $\mathfrak{p} = \mathfrak{p}^*$ is a minimal prime ideal of $(a) + \mathfrak{p}_0$, it follows that, in $R/\mathfrak{p}_0, \mathfrak{p}/\mathfrak{p}_0$ is a minimal prime ideal of the *principal ideal* $[(a) + \mathfrak{p}_0]/\mathfrak{p}_0$. But from $\mathfrak{p}/\mathfrak{p}_0 \supset \mathfrak{p}_1/\mathfrak{p}_0 \supset \mathfrak{p}_0/\mathfrak{p}_0$, we see that the rank of $\mathfrak{p}/\mathfrak{p}_0$ is at least two, and this contradicts Theorem 6.

COROLLARY. *Let $\mathfrak{p}$ be a proper prime ideal in a Noetherian ring and let l be a positive integer. Let $\mathfrak{p}'_1, \mathfrak{p}'_2, \ldots, \mathfrak{p}'_k$ be prime ideals, none of which contains $\mathfrak{p}$. If now there exists at least one chain $\mathfrak{p} = \mathfrak{p}_l \supset \mathfrak{p}_{l-1} \supset \ldots \supset \mathfrak{p}_1 \supset \mathfrak{p}_0$ of prime ideals, then such a chain can be found with the additional property that $\mathfrak{p}_1$ is not contained in any $\mathfrak{p}'_i$.*

Proof. We take any chain $\mathfrak{p} = \mathfrak{p}_l \supset \mathfrak{p}_{l-1} \supset \ldots \supset \mathfrak{p}_1 \supset \mathfrak{p}_0$. If $l = 1$ the assertion is trivial. If $l \geqslant 2$ we apply the proposition to $\mathfrak{p}_l$ and $\mathfrak{p}_{l-2}$ which shows us that we can replace $\mathfrak{p}_{l-1}$ by a prime ideal which is not contained in any $\mathfrak{p}'_i$. Having done this we apply the proposition to $\mathfrak{p}_{l-3}$ and to the new $\mathfrak{p}_{l-1}$. This shows us that we can now replace $\mathfrak{p}_{l-2}$ by a prime ideal which is not contained in any $\mathfrak{p}'_i$. Proceeding in this we can construct the required chain step by step.

The theorem which follows is an important generalization of Theorem 6.

THEOREM 7. *Let R be a Noetherian ring, let $(a_1, a_2, \ldots, a_m)$ be a proper ideal, and let $\mathfrak{p}$ be a minimal prime ideal of $(a_1, a_2, \ldots, a_m)$. Then the rank of $\mathfrak{p}$ is at most m.*

Proof. The proof will be by induction with respect to m. If $m = 1$ then the theorem has already been proved. In what follows, we suppose that $m \geqslant 2$, we suppose that the theorem has been proved for ideals with a basis of $m - 1$ elements, and we suppose ourselves confronted with the situation described in the statement of the theorem. Let $\mathfrak{p}'_1, \mathfrak{p}'_2, \ldots, \mathfrak{p}'_k$ be the minimal prime ideals of $(a_2, a_3, \ldots, a_m)$. By the inductive hypothesis, each of these has a rank not exceeding $m-1$, and therefore we may suppose that $\mathfrak{p}$ is not contained in any $\mathfrak{p}'_i$, because otherwise it is trivial that the rank of $\mathfrak{p}$ is not more than m. Suppose that we have a chain $\mathfrak{p} = \mathfrak{p}_l \supset \mathfrak{p}_{l-1} \supset \ldots \supset \mathfrak{p}_1 \supset \mathfrak{p}_0$ of prime ideals, where $l \geq 1$. We shall show that $l \leq m$. By the Corollary of Proposition 4, we may assume that $\mathfrak{p}_1$ is not contained in any $\mathfrak{p}'_i$, consequently, by Proposition 6 of §1·5, we can choose $b \in \mathfrak{p}_1$ so that b is not contained in any $\mathfrak{p}'_i$. Since $(b, a_2, \ldots, a_m) \subseteq \mathfrak{p}$, we can find a minimal prime ideal $\mathfrak{p}^*$ of $(b, a_2, \ldots, a_m)$ such that $\mathfrak{p}^* \subseteq \mathfrak{p}$. Since $\mathfrak{p}^* \supseteq (a_2, \ldots, a_m)$, we can find i so that $\mathfrak{p}^* \supseteq \mathfrak{p}'_i$, but by construction $\mathfrak{p}^* \neq \mathfrak{p}'_i$, and therefore $\mathfrak{p} \supseteq \mathfrak{p}^* \supset \mathfrak{p}'_i$. *We now assert that* $\mathfrak{p} = \mathfrak{p}^*$. For otherwise we should have $\mathfrak{p} \supset \mathfrak{p}^* \supset \mathfrak{p}'_i$, which shows that, in $R/(a_2, \ldots, a_m)$, $\mathfrak{p}/(a_2, \ldots, a_m)$ has rank at least equal to two. But $\mathfrak{p}/(a_2, \ldots, a_m)$ is a minimal prime ideal of the *principal ideal* $(a_1, a_2, \ldots, a_m)/(a_2, \ldots, a_m)$ and so we have a contradiction of Theorem 6. It has now been proved that $\mathfrak{p}$ is a minimal prime ideal of $(b, a_2, \ldots, a_m)$, and therefore that $\mathfrak{p}/(b)$ is a minimal prime ideal of $(b, a_2, \ldots, a_m)/(b)$ in $R/(b)$. But $(b, a_2, \ldots, a_m)/(b)$ has a base with $m-1$ elements, consequently, by our inductive hypothesis, the rank of $\mathfrak{p}/(b)$ is at most $m-1$. Now

$$\mathfrak{p}/(b) = \mathfrak{p}_l/(b) \supset \mathfrak{p}_{l-1}/(b) \supset \ldots \supset \mathfrak{p}_1/(b)$$

is a chain of prime ideals, and therefore it follows that $l-1 \leq m-1$. Thus $l \leq m$ and the theorem is proved.

Theorem 7 has a converse, namely: *If* $\mathfrak{p}$ *is a prime ideal (in a Noetherian ring) of rank* r $(r \geq 1)$ *then it is possible to find* r *elements* $a_1, a_2, \ldots, a_r$ *such that* $\mathfrak{p}$ *is a minimal prime ideal of* $(a_1, a_2, \ldots, a_r)$. The theorem which follows is a strengthened form of this converse.

THEOREM 8. *Suppose that* R *is Noetherian and that* $\mathfrak{a}$ *is a proper ideal of rank* r, *where* $r \geq 1$. *Then it is possible to find* r *elements* $a_1, a_2, \ldots, a_r$ *belonging to* $\mathfrak{a}$, *which are such that if* $1 \leq i \leq r$, *then* $(a_1, a_2, \ldots, a_i)$ *is of rank* i.

Proof. We shall build up the set $a_1, a_2, \ldots, a_r$ in stages by introducing one new element at each stage. Let us note that there are only a finite number of prime ideals of zero rank (because such a prime ideal must be a minimal prime ideal of (0)), and let us note, too, that none of these prime ideals can contain $\mathfrak{a}$, since, by hypothesis, rank $\mathfrak{a}=r\geqslant 1$. It is therefore possible (see Proposition 6 of § 1·5) to choose $a_1\in\mathfrak{a}$ so that a_1 is not in any prime ideal of zero rank. By construction rank $(a_1)\geqslant 1$, and, by Theorem 6, rank $(a_1)\leqslant 1$ consequently rank $(a_1)=1$. Let us suppose now that $a_1, a_2, \ldots, a_j$ have been constructed in accordance with the requirements of the theorem, and let us suppose, too, that $j<r$. Denote by $\mathfrak{p}_1, \mathfrak{p}_2, \ldots, \mathfrak{p}_h$ those prime ideals of $(a_1, a_2, \ldots, a_j)$ which are of rank j, and let $\mathfrak{p}_{h+1}, \mathfrak{p}_{h+2}, \ldots, \mathfrak{p}_k$ be the remaining prime ideals which belong to $(a_1, a_2, \ldots, a_j)$. We now have rank $\mathfrak{p}_\mu=j$ if $1\leqslant\mu\leqslant h$, and rank $\mathfrak{p}_\mu\geqslant j+1$ if $h<\mu\leqslant k$. Since rank $\mathfrak{a}=r>j$, $\mathfrak{a}$ cannot be contained in any of $\mathfrak{p}_1, \mathfrak{p}_2, \ldots, \mathfrak{p}_h$, consequently we can choose $a_{j+1}\in\mathfrak{a}$ so that a_{j+1} does not belong to any of $\mathfrak{p}_1, \mathfrak{p}_2, \ldots, \mathfrak{p}_h$. Let $\mathfrak{p}$ be any prime ideal of $(a_1, a_2, \ldots, a_j, a_{j+1})$, then we can find $\mu\leqslant k$ so that $\mathfrak{p}\supseteq\mathfrak{p}_\mu$. If $\mu>h$ then

$$\text{rank } \mathfrak{p}\geqslant\text{rank } \mathfrak{p}_\mu\geqslant j+1;$$

and if $\mu\leqslant h$ then, by construction, $\mathfrak{p}\neq\mathfrak{p}_\mu$ so that $\mathfrak{p}\supset\mathfrak{p}_\mu$ and rank $\mathfrak{p}>$ rank $\mathfrak{p}_\mu=j$. Thus, in any case, rank $\mathfrak{p}\geqslant j+1$ and therefore, since $\mathfrak{p}$ was any prime ideal of $(a_1, a_2, \ldots, a_{j+1})$, rank $(a_1, a_2, \ldots, a_{j+1})\geqslant j+1$. But by Theorem 7,

$$\text{rank } (a_1, a_2, \ldots, a_{j+1})\leqslant j+1,$$

consequently

$$\text{rank } (a_1, a_2, \ldots, a_{j+1})=j+1.$$

The existence of a full set of r elements, with the properties stated in the theorem, now follows by induction.

CHAPTER IV

THE ALGEBRAIC THEORY OF LOCAL RINGS

4·1. Notations and terminology. In this chapter and the next we shall make a systematic investigation of the properties of local rings. Usually we shall use Q to denote the local ring under consideration and $\mathfrak{m}$ will denote its maximal ideal. We put $P = Q/\mathfrak{m}$ so that P is a field. P will be termed the *residue field* of Q. We recall that, according to the Intersection Theorem, we have

$$\bigcap_{i=1}^{\infty} \mathfrak{m}^i = (0) \qquad (4\cdot1\cdot1)$$

for any local ring Q. Occasionally we shall wish to consider more than one local ring at the same time. In such a case we shall attach distinguishing marks to the symbol Q, and the same marks will be attached to the maximal ideal $\mathfrak{m}$ and the residue field P. For example, the maximal ideal of a local ring Q' will be written, automatically, as $\mathfrak{m}'$. By the *dimension*, $\dim Q$, of a local ring Q we shall mean the dimension of the zero ideal. Clearly $\dim Q = \operatorname{rank} \mathfrak{m}$, consequently, by Theorem 7 of § 3·5, the dimension of a local ring is finite and every base of $\mathfrak{m}$ contains at least $\dim Q$ elements. Note, first, that

$$\dim \mathfrak{p} + \operatorname{rank} \mathfrak{p} \leqslant \operatorname{rank} \mathfrak{m} = \dim Q \qquad (4\cdot1\cdot2)$$

for any prime ideal $\mathfrak{p}$; and, secondly, that *a local ring of dimension zero is the same as a primary ring* (§§ 3·3 and 3·4).

4·2. Systems of parameters. An important characterization of the dimension of a local ring is given by

THEOREM 1. *The dimension of a local ring Q is equal to the smallest number of non-zero elements required to generate an $\mathfrak{m}$-primary ideal.*

Proof. Let $\dim Q = d$. The case $d = 0$ is immediately settled because, then, Q is a primary ring and therefore (0) is an $\mathfrak{m}$-primary ideal. Suppose now that $d \geqslant 1$. If $(a_1, a_2, \ldots, a_s)$ is an

$\mathfrak{m}$-primary ideal, then, by Theorem 7 of §3·5, $d = \operatorname{rank}\mathfrak{m} \leqslant s$. Further, by Theorem 8 of §3·5, we can find d elements $b_1, b_2, \ldots, b_d$ contained in $\mathfrak{m}$, such that $(b_1, b_2, \ldots, b_d)$ has rank d. But $\mathfrak{m}$ is the only prime ideal of rank d, and no prime ideal has a larger rank, consequently $(b_1, b_2, \ldots, b_d)$ must be $\mathfrak{m}$-primary. This completes the proof.

DEFINITION. *If Q is a local ring of dimension $d \geqslant 1$, then a set of d elements which generates an $\mathfrak{m}$-primary ideal is called a 'system of parameters' in Q.*

The lemma which follows will be used in the proof of Corollary 2 of Theorem 2.

LEMMA 1. *Let R be a Noetherian ring and let $b \in R$, then b is a zero divisor if, and only if, b is contained in some prime ideal belonging to the zero ideal.*

Proof. The element b is a zero divisor if, and only if, $(0):(b) \neq (0)$. Lemma 1 now follows by applying Theorem 6 of §1·9.

THEOREM 2. *Let $\mathfrak{a} = (a_1, a_2, \ldots, a_s)$ be a proper ideal in the local ring Q, then $Q' = Q/\mathfrak{a}$ is a local ring and* $\dim Q \geqslant \dim Q' \geqslant \dim Q - s$. *Further,* $\dim Q' = \dim Q - s$ *if, and only if, $a_1, a_2, \ldots, a_s$ is a subset of a system of parameters.*

Proof. It is clear that every proper ideal of Q' is contained in $\mathfrak{m}' = \mathfrak{m}/\mathfrak{a}$, and therefore that Q' is a local ring. It is also clear that $\operatorname{rank}\mathfrak{m}' \leqslant \operatorname{rank}\mathfrak{m}$, that is to say, that $\dim Q' \leqslant \dim Q$. Let $t = \dim Q'$. If $t \geqslant 1$ we can find, by Theorem 1, elements $b_1, b_2, \ldots, b_t$ in Q, such that their residues modulo $\mathfrak{a}$ generate an $\mathfrak{m}'$-primary ideal, and then $(a_1, \ldots, a_s, b_1, \ldots, b_t)$ will be $\mathfrak{m}$-primary. By Theorem 7 of §3·5, we have $\dim Q = \operatorname{rank}\mathfrak{m} \leqslant s + t$ so that $\dim Q' \geqslant \dim Q - s$. If $t = 0$ then the zero ideal of Q' is $\mathfrak{m}'$-primary and therefore $(a_1, a_2, \ldots, a_s)$ must be $\mathfrak{m}$-primary. Thus $\dim Q = \operatorname{rank}\mathfrak{m} \leqslant s$ and again $\dim Q' \geqslant \dim Q - s$. If $\dim Q' = \dim Q - s$ then $s + t = \dim Q$, which shows that $(a_1, \ldots, a_s, b_1, \ldots, b_t)$ is a system of parameters. Finally, suppose that $(a_1, \ldots, a_s, c_1, \ldots, c_l)$ is a system of parameters. Let c_i' be the residue of c_i modulo $\mathfrak{a}$, then, since $(a_1, \ldots, a_s, c_1, \ldots, c_l)$ is $\mathfrak{m}$-primary, $(c_1', c_2', \ldots, c_l')$ is $\mathfrak{m}'$-primary, and therefore $\dim Q' \leqslant l$. Thus $\dim Q' \leqslant (l + s) - s = \dim Q - s$. But, in any case, $\dim Q' \geqslant \dim Q - s$ consequently $\dim Q' = \dim Q - s$ as required.

COROLLARY 1. *If $\mathfrak{a}$ is an ideal in the local ring Q then*

$$\bigcap_{i=1}^{\infty} (\mathfrak{a}+\mathfrak{m}^i) = \mathfrak{a}. \tag{4·2·1}$$

Proof. We may suppose that $\mathfrak{a} \neq (1)$. In that case $Q' = Q/\mathfrak{a}$ is a local ring and therefore $\bigcap_{i=1}^{\infty} (\mathfrak{m}')^i = (0)$. But $(\mathfrak{m}')^i = (\mathfrak{a}+\mathfrak{m}^i)/\mathfrak{a}$, hence, by Corollary 2 of Proposition 3 of § 2·2, the left-hand side of (4·2·1) corresponds to the zero ideal in Q' in the natural homomorphism of Q on to $Q/\mathfrak{a} = Q'$. This shows that the left-hand side of (4·2·1) is precisely $\mathfrak{a}$.

COROLLARY 2. *If $\mathfrak{a}$ is a proper ideal in the local ring Q, and if $\mathfrak{a}$ contains an element which is not a zero divisor, then*

$$\dim (Q/\mathfrak{a}) < \dim Q.$$

Further, if b is neither a unit nor a zero divisor then

$$\dim (Q/(b)) = \dim Q - 1.$$

Proof. If $\mathfrak{p}$ is a prime ideal containing $\mathfrak{a}$ then we can find another prime ideal $\mathfrak{p}_1$ such that $\mathfrak{p} \supset \mathfrak{p}_1$. For otherwise $\mathfrak{p}$ would be a minimal prime ideal of (0) and therefore, by Lemma 1, every element of $\mathfrak{a}$ would be a zero divisor. This shows that $\dim \mathfrak{p} < \operatorname{rank} \mathfrak{m}$, and therefore we have $\dim \mathfrak{a} < \operatorname{rank} \mathfrak{m} = \dim Q$. But $\dim (Q/\mathfrak{a}) = \dim \mathfrak{a}$ and so the first point is established. Putting $\mathfrak{a} = (b)$ we see that $\dim (Q/(b)) \leqslant \dim Q - 1$. The opposite inequality follows from Theorem 2 itself, and now the proof is complete.

The following proposition is extremely useful.

PROPOSITION 1. *Let $\mathfrak{a}$ and $\mathfrak{b}$ be ideals in a local ring Q which satisfy $\mathfrak{b} \subseteq \mathfrak{a} + \mathfrak{b}\mathfrak{m}$, then $\mathfrak{b} \subseteq \mathfrak{a}$.*

Proof. Since $\mathfrak{b} \subseteq \mathfrak{a} + \mathfrak{b}\mathfrak{m}$ we have $\mathfrak{b}\mathfrak{m} \subseteq \mathfrak{a}\mathfrak{m} + \mathfrak{b}\mathfrak{m}^2$ and therefore $\mathfrak{b} \subseteq \mathfrak{a} + \mathfrak{b}\mathfrak{m} \subseteq \mathfrak{a} + \mathfrak{b}\mathfrak{m}^2$. Again, $\mathfrak{b}\mathfrak{m} \subseteq \mathfrak{a}\mathfrak{m} + \mathfrak{b}\mathfrak{m}^3$, hence

$$\mathfrak{b} \subseteq \mathfrak{a} + \mathfrak{b}\mathfrak{m} \subseteq \mathfrak{a} + \mathfrak{b}\mathfrak{m}^3.$$

In this way we find that $\mathfrak{b} \subseteq \mathfrak{a} + \mathfrak{b}\mathfrak{m}^\nu$ for all ν, consequently, using (4·2·1), we have

$$\mathfrak{b} \subseteq \bigcap_{\nu=1}^{\infty} (\mathfrak{a} + \mathfrak{b}\mathfrak{m}^\nu) \subseteq \bigcap_{\nu=1}^{\infty} (\mathfrak{a} + \mathfrak{m}^\nu) = \mathfrak{a}.$$

4·3. Indeterminates. A system of parameters $t_1, t_2, \ldots, t_d$ has an important property which may be described, conveniently, by saying that the t_i are *analytically independent*. In proving that this property is possessed by a system of parameters, we shall introduce indeterminates in order to obtain elements which are, in a certain sense, non-special. We shall now prepare the way for this use of indeterminates by establishing some auxiliary results; however, so that we shall not make too great an interruption in the main argument, the proofs of two of these results (Propositions 2 and 5) will be postponed until § 4·11.

Throughout the remaining part of § 4·3, R will denote a Noetherian ring, and we shall put $R^* = R[z] = R[z_1, z_2, \ldots, z_p]$, where the z_i are indeterminates.† We recall that R^* will also be Noetherian (§ 1·10, Theorem 8, Corollary 1). Letters $\mathfrak{a}, \mathfrak{b}, \mathfrak{p}, \mathfrak{q}$, etc., will denote ideals in R and $\mathfrak{a}^*$, $\mathfrak{b}^*$, $\mathfrak{p}^*$, $\mathfrak{q}^*$, etc., will denote their respective extensions to R^*. An element of R^* may be written as $\phi(z)$ or as $\phi(z_1, z_2, \ldots, z_p)$, where ϕ is a polynomial with coefficients in R.

Lemma 2. *An element $\phi(z)$ of R^* will belong to $\mathfrak{a}^* = R^*\mathfrak{a}$ if, and only if, all the coefficients of ϕ are in $\mathfrak{a}$.*

Proof. If all the coefficients of ϕ are in $\mathfrak{a}$, then certainly $\phi(z) \in \mathfrak{a}^*$. If $\phi(z) \in \mathfrak{a}^*$, then $\phi(z) = a_1\psi_1(z) + a_2\psi_2(z) + \ldots + a_r\psi_r(z)$, where $a_i \in \mathfrak{a}$ and where the $\psi_i(z)$ are polynomials; and now, by comparing coefficients of each power-product of the z_ν, we see that all the coefficients of ϕ are in $\mathfrak{a}$.

Corollary 1. $\mathfrak{a}^* \cap R = \mathfrak{a}$.

Proof. We have $\mathfrak{a} \subseteq \mathfrak{a}^* \cap R$. Let $b \in \mathfrak{a}^* \cap R$. If we regard b as a 'constant polynomial' and apply the lemma we obtain $b \in \mathfrak{a}$.

Corollary 2. *If $\mathfrak{a} = \mathfrak{b}_1 \cap \mathfrak{b}_2 \cap \ldots \cap \mathfrak{b}_r$, then*

$$\mathfrak{a}^* = \mathfrak{b}_1^* \cap \mathfrak{b}_2^* \cap \ldots \cap \mathfrak{b}_r^*.$$

Proof. It is trivial that

$$\mathfrak{a}^* \subseteq \mathfrak{b}_1^* \cap \mathfrak{b}_2^* \cap \ldots \cap \mathfrak{b}_r^*.$$

Let
$$\phi(z) \in \mathfrak{b}_1^* \cap \mathfrak{b}_2^* \cap \ldots \cap \mathfrak{b}_r^*,$$

then for each i we have $\phi(z) \in \mathfrak{b}_i^*$ and therefore all the coefficients of $\phi(z)$ are in $\mathfrak{b}_i$. This shows that all the coefficients of $\phi(z)$ are in $\mathfrak{a}$, consequently $\phi(z) \in \mathfrak{a}^*$. This completes the proof.

† See 'Polynomial rings', § 1·10.

PROPOSITION 2. *If* $\mathfrak{q}$ *is* $\mathfrak{p}$*-primary then* $\mathfrak{p}^*$ *is prime and* $\mathfrak{q}^*$ *is* $\mathfrak{p}^*$*-primary.*

The proof of Proposition 2 is postponed until § 4·11.

PROPOSITION 3. *If* $\mathfrak{a} = \mathfrak{q}_1 \cap \mathfrak{q}_2 \cap \ldots \cap \mathfrak{q}_n$ *where* $\mathfrak{q}_i$ *is* $\mathfrak{p}_i$*-primary, then* $\mathfrak{a}^* = \mathfrak{q}_1^* \cap \mathfrak{q}_2^* \cap \ldots \cap \mathfrak{q}_n^*$, *where* $\mathfrak{q}_i^*$ *is* $\mathfrak{p}_i^*$*-primary. Further, if the first decomposition is normal so is the second.*

Proof. The first assertion follows at once from Proposition 2 and Corollary 2 of Lemma 2. Now assume that $\mathfrak{a} = \mathfrak{q}_1 \cap \mathfrak{q}_2 \cap \ldots \cap \mathfrak{q}_n$ is a normal decomposition. Since $\mathfrak{p}_i^* \cap R = \mathfrak{p}_i$ (Lemma 2, Corollary 1), and since the $\mathfrak{p}_i$ are distinct, it follows that the $\mathfrak{p}_i^*$ are distinct. Further, the decomposition $\mathfrak{a}^* = \mathfrak{q}_1^* \cap \mathfrak{q}_2^* \cap \ldots \cap \mathfrak{q}_n^*$ is irredundant. For suppose, for example, that $\mathfrak{q}_1^* \supseteq \mathfrak{q}_2^* \cap \ldots \cap \mathfrak{q}_n^*$, then by projecting upon R and using Corollary 1 of Lemma 2 we at once obtain a contradiction. This completes the proof.

PROPOSITION 4. *If* $\phi(z)$ *is a zero divisor in* R^*, *then we can find* $c \neq 0$ *in* R *such that* $c\phi(z) = 0$.

Proof. By Lemma 1, $\phi(z) \in \mathfrak{p}^*$, where $\mathfrak{p}^*$ is a prime ideal belonging to the zero ideal in R^*. By applying Proposition 3 to the zero ideal of R, we see that $\mathfrak{p}^*$ is the extension of a prime ideal $\mathfrak{p}$ belonging to the ideal (0) in R. Since $\phi(z) \in \mathfrak{p}^*$ it follows, by Lemma 2, that all the coefficients of $\phi(z)$ are in $\mathfrak{p}$. But $(0) : \mathfrak{p} \neq (0)$ (§ 1·9, Theorem 6), consequently we can find $c \neq 0$ such that $c \in (0) : \mathfrak{p}$. We then have $c\phi(z) = 0$.

PROPOSITION 5. *If* $\mathfrak{p}$ *is a prime ideal then* rank $\mathfrak{p}$ = rank $\mathfrak{p}^*$.

The proof of Proposition 5 will be given in § 4·11.

4·4. Analytic independence. Let $\phi(z_1, z_2, \ldots, z_r)$ be a polynomial in r variables with coefficients in a local ring Q, then the polynomial ϕ will be called *a form of degree s* ($s \geqslant 0$) if all the non-zero terms are of degree s. We can now define *analytic independence* with the aid of this concept.

DEFINITION. *Let* $t_1, t_2, \ldots, t_r$ *be elements of* $\mathfrak{m}$, *then the* t_i *will be said to be 'analytically independent' if the following condition is satisfied: whenever* $\phi(t_1, t_2, \ldots, t_r) = 0$, *where* $\phi(z_1, z_2, \ldots, z_r)$ *is a form of arbitrary degree, then all the coefficients of* ϕ *are in* $\mathfrak{m}$.

Note that *if* $t_1, t_2, \ldots, t_r$ *are analytically independent, if* $\psi(z_1, z_2, \ldots, z_r)$ *is a form of degree* $s \geqslant 0$, *and if not all the coefficients of* ψ *are in* $\mathfrak{m}$, *then* $\psi(t_1, t_2, \ldots, t_r) \notin \mathfrak{m}(t_1, t_2, \ldots, t_r)^s$. For if $\psi(t) = \psi(t_1, t_2, \ldots, t_r) \in \mathfrak{m}(t_1, t_2, \ldots, t_r)^s$, then $\psi(t) = \psi_0(t)$, where ψ_0 is a form of degree s with coefficients in $\mathfrak{m}$. Thus $\phi(t) = 0$, where $\phi = \psi - \psi_0$, hence, since the t_i are analytically independent, all the coefficients of ϕ are in $\mathfrak{m}$. But this requires that all the coefficients of $\psi = \phi + \psi_0$ be in $\mathfrak{m}$, and therefore we have a contradiction.

THEOREM 3. *If* $t_1, t_2, \ldots, t_d$ *is a system of parameters in a local ring* Q, *then the* t_i *are analytically independent.*

Proof. Let $\phi(t) = \phi(t_1, t_2, \ldots, t_d) = 0$, where ϕ is a form of degree s, then we have to show that all the coefficients of ϕ are in $\mathfrak{m}$. Let c be the coefficient of t_1^s in ϕ. *We shall show, first, that* $c \in \mathfrak{m}$. We have $ct_1^s \in (t_2, \ldots, t_d)$. If $c \notin \mathfrak{m}$ then c is a unit and $t_1^s \in (t_2, \ldots, t_d)$. Since $t_1^s \in (t_2, \ldots, t_d)$ we have $(t_1, t_2, \ldots, t_d)^s \subseteq (t_2, \ldots, t_d)$, and hence, because $(t_1, t_2, \ldots, t_d)$ is $\mathfrak{m}$-primary, $(t_2, \ldots, t_d)$ contains a power of $\mathfrak{m}$ and therefore $(t_2, \ldots, t_d)$ is also $\mathfrak{m}$-primary. This shows that $\dim Q \leqslant d - 1$; but, $t_1, t_2, \ldots, t_d$ being a system of parameters, $\dim Q = d$. This contradiction establishes that $c \in \mathfrak{m}$.

In order to extend this result to *all* the coefficients of ϕ, we introduce d^2 indeterminates z_{ij} $(1 \leqslant i \leqslant d, 1 \leqslant j \leqslant d)$ and put $R^* = Q[z_{ij}]$, $\mathfrak{m}^* = R^*\mathfrak{m}$. By Proposition 2, $\mathfrak{m}^*$ is a prime ideal. We assert that *no element of* $R^* - \mathfrak{m}^*$ *is a zero divisor in* R^*. For let $\psi(z_{ij}) \in R^*$ and $\psi(z_{ij}) \notin \mathfrak{m}^*$, then there is a coefficient of ψ which is not in $\mathfrak{m}$, and such a coefficient will be a unit in Q. It now follows from Proposition 4 that $\psi(z_{ij})$ cannot be a zero divisor in R^*. Since no element in $R^* - \mathfrak{m}^*$ is a zero divisor, we can form an ordinary ring of quotients Q' of R^* with respect to $\mathfrak{m}^*$. Q' is a local ring and $\mathfrak{m}' = Q'\mathfrak{m}^* = Q'\mathfrak{m}$ is its maximal ideal; further, using Proposition 5, rank $\mathfrak{m}' = \operatorname{rank} \mathfrak{m}^* = \operatorname{rank} \mathfrak{m}$, which shows that $\dim Q' = \dim Q = d$. Again, $R^*t_1 + R^*t_2 + \ldots + R^*t_d$ is $\mathfrak{m}^*$-primary (Proposition 2), consequently $Q't_1 + Q't_2 + \ldots + Q't_d$ is $\mathfrak{m}'$-primary, which shows that $t_1, t_2, \ldots, t_d$ is a system of parameters in Q'.

Consider the determinant $|z_{ij}|$. By Lemma 2, $|z_{ij}| \notin \mathfrak{m}^*$, consequently $|z_{ij}|$ is a unit in Q'. We can therefore define $u_1, u_2, \ldots, u_d$ by $t_i = \Sigma_j z_{ij} u_j$, and then

$$Q'u_1 + Q'u_2 + \ldots + Q'u_d = Q't_1 + Q't_2 + \ldots + Q't_d,$$

so that $u_1, u_2, \ldots, u_d$ is a system of parameters in Q'. Now

$$\phi(t) = \phi(\Sigma_j z_{1j} u_j, \Sigma_j z_{2j} u_j, \ldots, \Sigma_j z_{dj} u_j) = f(u),$$

where f is a form of degree s in the u_j, which is to be obtained from ϕ in the obvious manner. But $f(u) = \phi(t) = 0$, consequently (by the first part) the coefficient of u_1^s, that is, $\phi(z_{11}, z_{21}, \ldots, z_{d1})$, is in $\mathfrak{m}'$. Thus $\phi(z_{11}, z_{21}, \ldots, z_{d1}) \in \mathfrak{m}' \cap R^* = \mathfrak{m}^*$, hence all the coefficients of $\phi(z_{11}, z_{21}, \ldots, z_{d1})$ are in $\mathfrak{m}$ (Lemma 2). This, however, is exactly what we had to prove.

4·5. Mimimal ideal bases. Let $\mathfrak{a} \neq (0)$ be a proper ideal in a local ring Q. Let us make the

DEFINITION. *If $\mathfrak{a} = (a_1, a_2, \ldots, a_s)$, and if no proper subset of the a_i generates $\mathfrak{a}$, then we shall say that $a_1, a_2, \ldots, a_s$ is a 'minimal base' of $\mathfrak{a}$.*

The elements in a minimal base will all be different from zero, accordingly, we shall make a conventional extension of the definition and say that the ideal (0) has no minimal bases. We shall also say that the number of elements in a minimal base of (0) is zero. Suppose that $\mathfrak{a}$ is a proper ideal (not excluding the zero ideal), then, by Proposition 2 of § 3·3, $\mathfrak{a}/\mathfrak{a}\mathfrak{m}$ is a vector space over the residue field P; we denote the dimension of this vector space by $\dim_P(\mathfrak{a}/\mathfrak{a}\mathfrak{m})$. For the purposes of Proposition 6, $\bar{a}$ will be used to denote the residue modulo $\mathfrak{a}\mathfrak{m}$ of an element a of $\mathfrak{a}$.

PROPOSITION 6. *Let $a_i \in \mathfrak{a}$ for $1 \leqslant i \leqslant s$, then $\mathfrak{a} = (a_1, a_2, \ldots, a_s)$ if, and only if, $\mathfrak{a}/\mathfrak{a}\mathfrak{m} = P\bar{a}_1 + P\bar{a}_2 + \ldots + P\bar{a}_s$. Further, the a_i form a minimal base of $\mathfrak{a}$ if, and only if, the $\bar{a}_i$ form a linearly independent base of $\mathfrak{a}/\mathfrak{a}\mathfrak{m}$ over P; and every minimal base of $\mathfrak{a}$ contains exactly $\dim_P(\mathfrak{a}/\mathfrak{a}\mathfrak{m})$ elements.*

Proof. It is clear from the definitions that if $\mathfrak{a} = (a_1, a_2, \ldots, a_s)$, then $\mathfrak{a}/\mathfrak{a}\mathfrak{m} = P\bar{a}_1 + P\bar{a}_2 + \ldots + P\bar{a}_s$. Assume now that

$$\mathfrak{a}/\mathfrak{a}\mathfrak{m} = P\bar{a}_1 + P\bar{a}_2 + \ldots + P\bar{a}_s$$

and let $a \in \mathfrak{a}$. We can then find elements $\rho_i \in P$ such that

$$\bar{a} = \rho_1 \bar{a}_1 + \rho_2 \bar{a}_2 + \ldots + \rho_s \bar{a}_s.$$

Let $r_i \in R$ be a representative of ρ_i, then

$$a \equiv r_1 a_1 + r_2 a_2 + \ldots + r_s a_s$$

modulo $\mathfrak{a}\mathfrak{m}$, consequently $a \in (a_1, a_2, \ldots, a_s) + \mathfrak{a}\mathfrak{m}$. But, since a was any element of $\mathfrak{a}$, this means that $\mathfrak{a} \subseteq (a_1, a_2, \ldots, a_s) + \mathfrak{a}\mathfrak{m}$, hence, by Proposition 1 and our hypotheses, $\mathfrak{a} \subseteq (a_1, a_2, \ldots, a_s) \subseteq \mathfrak{a}$. This proves the first assertion. The second assertion follows immediately from the first, and the third assertion follows immediately from the second.

4·6. Regular local rings. The dimension of a local ring Q is not greater than the number of elements in a minimal base of $\mathfrak{m}$, consequently (Proposition 6)

$$\dim Q \leqslant \dim_P (\mathfrak{m}/\mathfrak{m}^2). \tag{4·6·1}$$

A local ring for which the condition

$$\dim Q = \dim_P (\mathfrak{m}/\mathfrak{m}^2) \tag{4·6·2}$$

is satisfied is said to be *regular*. By Proposition 6, *Q is regular if, and only if, the number of elements in a minimal base of $\mathfrak{m}$ is equal to the dimension of Q*; hence, *if Q is regular, then the elements of a minimal base of $\mathfrak{m}$ form a system of parameters.* Note that if Q is regular and of dimension zero, we have $\mathfrak{m} = \mathfrak{m}^2$, and therefore $\mathfrak{m} = \mathfrak{m}^s$ for all values of s. But, since the intersection of all the powers $\mathfrak{m}^s$ is the zero ideal, this means that $\mathfrak{m} = (0)$; hence *a regular local ring of dimension zero is the same as a field.* Regular local rings are important in the applications of ideal theory to geometry, because the local ring of a simple point on an irreducible variety is always regular.

We observed above that, in the case of a regular ring, the elements of a minimal base of $\mathfrak{m}$ form a system of parameters, consequently they must be analytically independent (Theorem 3). It will now be shown that the converse is true, namely, that if the elements of a minimal base of $\mathfrak{m}$ are analytically independent then the ring must be regular. To facilitate the proof of this result we shall suppose (until Lemmas 3 and 4 have been established) that Q is a given local ring, that $(u_1, u_2, \ldots, u_n)$ is a fixed minimal base of $\mathfrak{m}$, and that the u_i are analytically independent.

LEMMA 3. *Suppose that $\alpha \in \mathfrak{m}^h$, $\alpha \notin \mathfrak{m}^{h+1}$, $\beta \in \mathfrak{m}^k$, $\beta \notin \mathfrak{m}^{k+1}$ (where h and k are non-negative integers†) then $\alpha\beta \notin \mathfrak{m}^{h+k+1}$.*

† By $\mathfrak{m}^0$ we mean, of course the whole ring Q.

Proof. Since $\alpha \in \mathfrak{m}^h = (u_1, u_2, \ldots, u_n)^h$ we have

$$\alpha = \phi(u_1, u_2, \ldots, u_n) = \phi(u),$$

where ϕ is a form of degree h; and, since $\alpha \notin \mathfrak{m}^{h+1}$, it follows that not all the coefficients of ϕ are in $\mathfrak{m}$. Similarly, $\beta = \psi(u)$, where ψ is a form of degree k and not all the coefficients of ψ are in $\mathfrak{m}$. Let $\chi = \phi\psi$ (so that χ is a form of degree $h+k$), let $z_1, z_2, \ldots, z_n$ be indeterminates, let $R^* = Q[z]$, and let $\mathfrak{m}^* = R^*\mathfrak{m}$. We then have $\phi(z) \notin \mathfrak{m}^*$, $\psi(z) \notin \mathfrak{m}^*$ (Lemma 2), hence, since $\mathfrak{m}^*$ is prime (Proposition 2), $\chi(z) = \phi(z)\psi(z) \notin \mathfrak{m}^*$. Thus, by Lemma 2, not all the coefficients of χ are in $\mathfrak{m}$, consequently (since by hypothesis the u_i are analytically independent) $\alpha\beta = \chi(u) \notin \mathfrak{m}^{h+k+1}$.

COROLLARY. *Q is an integral domain.*

For if $\alpha \neq 0$ and $\beta \neq 0$ we can find integers h and k such that the conditions of Lemma 3 are satisfied, and then $\alpha\beta \notin \mathfrak{m}^{h+k+1}$ so that *a fortiori* $\alpha\beta \neq 0$.

LEMMA 4. *Suppose that $n \geq 2$, put $Q' = Q/(u_1)$, and let v_i $(2 \leq i \leq n)$ be the residue of u_i modulo (u_1). Then $\mathfrak{m}' = (v_2, \ldots, v_n)$, and this base is minimal. Further, $v_2, \ldots, v_n$ are analytically independent in Q'.*

Proof. It is clear that $v_2, \ldots, v_n$ form a minimal base of $\mathfrak{m}'$. Let $\phi'(v_2, \ldots, v_n) = 0$, where ϕ' is a form of degree s with coefficients in Q', then it is sufficient to prove that all the coefficients of ϕ' are in $\mathfrak{m}'$. Let ϕ be a form of degree s obtained by replacing each coefficient of ϕ' by one of its representatives, then the proof will be complete if we show that all the coefficients of ϕ are in $\mathfrak{m}$. Assume the contrary, then

$$\phi(u_2, \ldots, u_n) \in \mathfrak{m}^s$$

but
$$\phi(u_2, \ldots, u_n) \notin \mathfrak{m}^{s+1}$$

Again, since
$$\phi'(v_2, \ldots, v_n) = 0,$$

we have
$$\phi(u_2, \ldots, u_n) \in (u_1),$$

say
$$\phi(u_2, \ldots, u_n) = \alpha u_1.$$

The element α is not zero because

$$\phi(u_2, \ldots, u_n) \notin \mathfrak{m}^{s+1};$$

therefore, if we choose h so that $\alpha \in \mathfrak{m}^h, \alpha \notin \mathfrak{m}^{h+1}$, then we can write

$$\alpha = \psi(u_1, u_2, \ldots, u_n),$$

where ψ is a form of degree h and not all the coefficients of ψ are in $\mathfrak{m}$. It follows that

$$\alpha u_1 = u_1 \psi(u_1, u_2, \ldots, u_n)$$

belongs to $\mathfrak{m}^{h+1}$ but not to $\mathfrak{m}^{h+2}$, consequently $h+1=s$. Let $z_1, z_2, \ldots, z_n$ be indeterminates and let us define a form χ of degree s by

$$\chi(z) = \chi(z_1, z_2, \ldots, z_n) = \phi(z_2, \ldots, z_n) - z_1 \psi(z_1, z_2, \ldots, z_n).$$

Not all the coefficients of χ are in $\mathfrak{m}$, hence, since the u_i are analytically independent, $\chi(u) \notin \mathfrak{m}^{s+1}$. But, by construction, $\chi(u) = 0$. This contradiction establishes the lemma.

COROLLARY 1. *The ideals* (0), (u_1), (u_1, u_2), (u_1, u_2, u_3), $\ldots$, $(u_1, u_2, \ldots, u_n)$ *are all prime.*

Proof. The assertion that (0) is a prime ideal is equivalent to the assertion that Q is an integral domain, and this has already been proved (Lemma 3, Corollary). Since, by Lemma 4, $Q' = Q/(u_1)$ has the same properties as were assumed to hold for Q, Q' is also an integral domain, and therefore (u_1) is a prime ideal (§2·3, Proposition 4). This latter result applied to Q' shows that $(v_2) = (u_1, u_2)/(u_1)$ is a prime ideal, consequently (u_1, u_2), too, is prime. Proceeding in this way we establish the corollary.

COROLLARY 2. *Q is a regular local ring.*

Proof. It is enough to show that $\dim Q = n$ (the number of elements in a minimal base of $\mathfrak{m}$), and we know, without making use of our special assumptions, that $\dim Q \leqslant n$. But $\dim Q \geqslant n$, because, as we saw in Corollary 1,

$$(0) \subset (u_1) \subset (u_1, u_2) \subset (u_1, u_2, u_3) \subset \ldots \subset (u_1, u_2, \ldots, u_n) = \mathfrak{m}$$

is a chain of prime ideals.

These results will now be put into the form of theorems.

THEOREM 4. *Let Q be a local ring and let $(u_1, u_2, \ldots, u_n)$ be a minimal base of $\mathfrak{m}$, then Q is regular if, and only if, the u_i are analytically independent.*

THEOREM 5. *Every regular local ring is an integral domain.*

This follows by combining Theorem 4 with the Corollary of Lemma 3.

THEOREM 6. *If Q is a regular local ring and $(u_1, u_2, \ldots, u_n)$ is a minimal base of $\mathfrak{m}$, then $(u_1, u_2, \ldots, u_i)$ is a prime ideal for $1 \leqslant i \leqslant n$. Further, $Q/(u_1, u_2, \ldots, u_i)$ is a regular local ring of dimension $n-i$.*

Proof. The first assertion follows by combining Theorem 4 with Corollary 1 of Lemma 4. Put $Q^* = Q/(u_1, u_2, \ldots, u_i)$. By Theorem 2, $\dim Q^* \geqslant n-i$, but, since

$$(u_1, u_2, \ldots, u_i) \subset (u_1, u_2, \ldots, u_i, u_{i+1}) \subset \ldots \subset (u_1, u_2, \ldots, u_n)$$

is a chain of prime ideals we have the opposite inequality, and therefore $\dim Q^* = n-i$. Now $\mathfrak{m}^* = (u_1, u_2, \ldots, u_n)/(u_1, u_2, \ldots, u_i)$ has a minimal base of $n-i$ elements, namely, the residues of $u_{i+1}, u_{i+2}, \ldots, u_n$ modulo $(u_1, u_2, \ldots, u_i)$, hence Q^* must be regular.

4·7. Integral dependence and integral closure. For the remainder of this chapter we shall, to a great extent, confine our attention to one-dimensional regular local rings and their application to the so-called *Theory of Divisors*. Two new concepts, namely, the concepts of *integral dependence* and *integral closure*, will now be introduced because we shall require them in our investigations. Let R' be an extension ring of a ring R, then an element $\alpha \in R'$ is said to be *integral with respect to R*, if an equation of the form

$$\alpha^n + a_1\alpha^{n-1} + a_2\alpha^{n-2} + \ldots + a_n = 0$$

holds, where the a_i are in R.

DEFINITION. *If every element of R', which is integral with respect to R, belongs to R then R is said to be 'integrally closed' in R'.*

A particularly important situation arises when we have a ring which is integrally closed in its full ring of quotients.

THEOREM 7. *Let R be a Noetherian ring which is integrally closed in its full ring $\mathfrak{R}$ of quotients; let α be an element of R which is neither a unit nor a zero divisor in R; and let $\mathfrak{p}$ be any prime ideal which belongs to (α). In these circumstances, $\mathfrak{p}$ is of rank one, and if Q is the (generalized) ring of quotients of R with respect to $\mathfrak{p}$ then Q is a regular local ring of dimension one.*

Proof. Since $\mathfrak{p}$ belongs to (α), we have, by Theorem 6 of § 1·9, $(\alpha) \subset (\alpha):\mathfrak{p}$, where the inclusion is strict, consequently we can choose $\beta \in (\alpha):\mathfrak{p}$ so that $\beta \notin (\alpha)$. By hypothesis, α is not a zero divisor, hence $\zeta = \beta/\alpha$ exists as an element of $\mathfrak{R}$. *This element ζ does not belong to R because $\beta \notin (\alpha)$.* Further, since $\beta\mathfrak{p} \subseteq (\alpha)$ we have $\zeta\mathfrak{p} \subseteq R$, which shows that $\zeta\mathfrak{p}$ is an ideal of R. *We claim that $\zeta\mathfrak{p} \nsubseteq \mathfrak{p}$.* For assume that $\zeta\mathfrak{p} \subseteq \mathfrak{p}$ and let $\mathfrak{p} = (u_1, u_2, \ldots, u_n)$, then for each i, from 1 to n inclusive, we have $\zeta u_i = \sum_{j=1}^{n} r_{ij} u_j$ with suitable elements $r_{ij} \in R$. If now δ_{ij} has its usual meaning, this can be written as $\sum_j (\zeta\delta_{ij} - r_{ij})\, u_j = 0$, from which it follows that $\Delta u_j = 0$ (for $1 \leqslant j \leqslant n$), where Δ is the determinant $|\,\zeta\delta_{ij} - r_{ij}\,|$. We now see that $\Delta\mathfrak{p} = \Delta(u_1, u_2, \ldots, u_n) = (0)$, so that, in particular, $\Delta\alpha = 0$; consequently, since α is not a zero divisor, $\Delta = 0$. But $\Delta = 0$ implies that we have an equation of the form

$$\zeta^n + a_1\zeta^{n-1} + a_2\zeta^{n-2} + \ldots + a_n = 0,$$

where the a_i are in R, or, in other words, $\Delta = 0$ implies that ζ is integral with respect to R. Since R is integrally closed in $\mathfrak{R}$, this means that $\zeta \in R$, and so we have a contradiction.

It has now been proved that $\zeta\mathfrak{p} \nsubseteq \mathfrak{p}$. Choose $p \in \mathfrak{p}$ so that $c = \zeta p \notin \mathfrak{p}$, then $\beta p = \alpha c$ and $c \notin \mathfrak{p}$. Let us now turn our attention to the ring Q of quotients of R with respect to $\mathfrak{p}$. Let $S = R - \mathfrak{p}$, let $\mathfrak{n}$ be the S-component of the zero ideal, let $R^* = R/\mathfrak{n}$, let $\mathfrak{p}^* = \mathfrak{p}/\mathfrak{n}$, and let σ be the natural homomorphism of R on to $R/\mathfrak{n} = R^*$. Q is now the ordinary ring of quotients of R^* with respect to $\mathfrak{p}^*$. Q is, of course, a local ring, and we shall (as usual) use $\mathfrak{m}$ to denote its maximal ideal. Put $\sigma(\alpha) = \alpha^*$, $\sigma(\beta) = \beta^*$, $\sigma(p) = p^*$ and $\sigma(c) = c^*$, then $\beta^* p^* = \alpha^* c^*$ and

$$Q\alpha^* c^* = Q\beta^* p^* \subseteq \beta^*\mathfrak{m}.$$

Now $c \notin \mathfrak{p}$, and this shows that $c^* \notin \mathfrak{p}^*$, consequently c^* is a unit in Q and therefore we have

$$Q\alpha^* = Q\alpha^* c^* \subseteq \beta^*\mathfrak{m}.$$

Again, $\beta\mathfrak{p} \subseteq (\alpha)$ and therefore $\beta^*\mathfrak{p}^* \subseteq R^*\alpha^*$, which shows that $\beta^*\mathfrak{m} \subseteq Q\alpha^*$. Combining these results we obtain $\beta^*\mathfrak{m} = Q\alpha^*$. Choose $\theta \in \mathfrak{m}$ so that $\alpha^* = \beta^*\theta$; we then have $\beta^*\mathfrak{m} = \beta^* Q\theta$, and from this we shall obtain $\mathfrak{m} = Q\theta$ by showing that β^* is not a zero divisor in Q.

The assertion that β^* is not a zero divisor in Q will be proved if we show that β^* is not a zero divisor in R^*. Assume that

$\beta^* z^* = 0$, where $z \in R$ and $z^* = \sigma(z)$, then $\beta z \in \mathfrak{n}$, and therefore we can find $c_0 \notin \mathfrak{p}$ such that $c_0 \beta z = 0$. If now we multiply the equation $\beta p = \alpha c$ by $c_0 z$, we find that $0 = \alpha c c_0 z$, hence, since α is not a zero divisor, $0 = c c_0 z$. But $c c_0 \notin \mathfrak{p}$, consequently $z \in \mathfrak{n}$ and $z^* = 0$. It is now established that β^* is not a zero divisor and this, as has already been observed, allows us to conclude that $\mathfrak{m} = Q\theta$.

To complete the proof, we recall that $\mathfrak{p}$ contains α which is not a zero divisor, consequently $\mathfrak{p}$ does not belong to the zero ideal (see Lemma 1), and this shows that rank $\mathfrak{p} \geqslant 1$. By Proposition 3 of § 3·5, rank $\mathfrak{p}$ = rank $\mathfrak{m}$, hence $\dim Q = \text{rank}\, \mathfrak{m} = \text{rank}\, \mathfrak{p} \geqslant 1$. But $\mathfrak{m} = Q\theta$, consequently, by Theorem 6 of § 3·5, rank $\mathfrak{m} \leqslant 1$, and so we must have $\dim Q = \text{rank}\, \mathfrak{p} = 1$. Finally, (θ) is a minimal base of $\mathfrak{m}$ because, since rank $\mathfrak{m} = 1$, $\mathfrak{m} \neq (0)$ and this shows that Q is regular.

4·8. One-dimensional regular local rings. The ideal theory of a one-dimensional regular local ring is extremely simple, because, as we shall now show, every proper ideal in such a ring (with the exception of the zero ideal) is a power of the maximal ideal.

PROPOSITION 7. *Let Q be a regular local ring of dimension one and let $\mathfrak{a}$ be a non-zero ideal. Then $\mathfrak{m}$ is a principal ideal, say $\mathfrak{m} = (\pi)$, and with a suitable non-negative integer μ we have $\mathfrak{a} = \mathfrak{m}^\mu = (\pi^\mu)$.*

Proof. The maximal ideal $\mathfrak{m}$ is principal, because, since Q is regular and of dimension one, a minimal base of $\mathfrak{m}$ must contain only a single element. Let $\mathfrak{m} = (\pi)$ and choose $\mu \geqslant 0$ so that $\mathfrak{a} \subseteq \mathfrak{m}^\mu$ but $\mathfrak{a} \nsubseteq \mathfrak{m}^{\mu+1}$; we can then find $a \in \mathfrak{a}$ so that $a \notin \mathfrak{m}^{\mu+1} = (\pi^{\mu+1})$. Now $a \in (\pi^\mu)$, say $a = \pi^\mu u$, and u cannot belong to $\mathfrak{m} = (\pi)$ because $a \notin (\pi^{\mu+1})$. Thus u is a unit and $(\pi^\mu) = (a) \subseteq \mathfrak{a} \subseteq (\pi^\mu)$. The proof is now complete.

Let the hypotheses be as in Proposition 7. By Theorem 5, Q is an integral domain, consequently it has a quotient field F (say). Suppose that $\alpha \in F$ ($\alpha \neq 0$), then $\alpha = a/b$, where $a \in Q$ and $b \in Q$. Proposition 7 shows that, with suitable integers μ and ν, $(a) = (\pi^\mu)$ and $(b) = (\pi^\nu)$, consequently $a = \pi^\mu u_1$ and $b = \pi^\nu u_2$, where u_1 and u_2 are units in Q. It follows that $\alpha = \pi^\rho u$, where ρ is an integer (which may be positive, negative, or zero) and where u is a unit in Q. These remarks will be used in the proof of Theorem 8, which gives a useful criterion for deciding when a one-dimensional local ring is regular.

THEOREM 8. *Let Q be a local ring of dimension one, then, in order that Q shall be regular, it is necessary and sufficient that Q be an integral domain which is integrally closed in its quotient field.*

Proof. First, let us assume that Q is regular. We know, then, that Q is an integral domain. Let F be the quotient field of Q, and let α be an element of F which is integral with respect to Q. We wish to prove that α belongs to Q. Assume the contrary, then $\alpha \neq 0$, consequently we can write α in the form $\alpha = \pi^\rho u$, where $\mathfrak{m} = (\pi)$, where u is a unit in Q, and where ρ is an integer. Since $\alpha \notin Q$, ρ must be negative, say $\rho = -\lambda$ where $\lambda > 0$. Further, since $\alpha = \pi^\rho u$ is integral with respect to Q, an equation of the form

$$(\pi^\rho u)^n + a_1(\pi^\rho u)^{n-1} + a_2(\pi^\rho u)^{n-2} + \dots + a_n = 0$$

holds, where the a_i are in Q. Multiplying this equation by $\pi^{\lambda n}$, we obtain $u^n + r\pi = 0$, where $r \in Q$; thus $u^n \in \mathfrak{m}$, and now we have a contradiction because u is a unit.

We have now to establish the converse, so we suppose that Q is an integral domain which is integrally closed in its quotient field F. Since $\dim Q = 1$, $\mathfrak{m} \neq (0)$; we can therefore choose $\alpha \in \mathfrak{m}$ so that $\alpha \neq 0$. But Q is an integral domain, consequently (α) has rank one and therefore (α) is $\mathfrak{m}$-primary. It follows *a fortiori* that $\mathfrak{m}$ belongs to (α). If we now apply Theorem 7 (at the same time observing that the ring of quotients of Q with respect to $\mathfrak{m}$ is Q itself) we see that Q is regular.

4·9. Divisors. Throughout the whole of §4·9 we shall be concerned with a fixed Noetherian ring R which is integrally closed in its full ring $\mathfrak{R}$ of quotients. We shall call a prime ideal $\mathfrak{p}$ of R a *relevant* prime ideal if, first, $\mathfrak{p}$ is of rank one and, secondly, at least one element of $\mathfrak{p}$ is not a zero divisor. Suppose that $\mathfrak{p}$ is a relevant prime ideal and let b be any element of $\mathfrak{p}$ which is not a zero divisor, then (b) is of rank one and therefore, since rank $\mathfrak{p} = 1$, $\mathfrak{p}$ is a minimal prime ideal of (b). Further, we can apply Theorem 7 and so obtain the following result: *if $\mathfrak{p}$ is a relevant prime ideal of R then the ring of quotients of R with respect to $\mathfrak{p}$ is a regular local ring of dimension one.*

PROPOSITION 8. *Let $\mathfrak{p}$ be a relevant prime ideal of R, then every primary ideal which belongs to $\mathfrak{p}$ is a symbolic prime power of $\mathfrak{p}$.*

Proof. Let Q be the ring of quotients of R with respect to $\mathfrak{p}$ and let $\mathfrak{q}$ be a $\mathfrak{p}$-primary ideal. Choose an integer s so that $\mathfrak{p}^s \subseteq \mathfrak{q}$, then $Q\mathfrak{q} \supseteq Q\mathfrak{p}^s = \mathfrak{m}^s$, where $\mathfrak{m}$ is the maximal ideal of Q. (We recall that $Q\mathfrak{q}$ is a 'conventional extension' in the sense which was explained in § 2·7). Now Q is regular and of dimension one and $Q\mathfrak{q} \supseteq \mathfrak{m}^s$ implies that $Q\mathfrak{q} \neq (0)$, consequently, by Proposition 7, $Q\mathfrak{q} = \mathfrak{m}^r = Q\mathfrak{p}^r$ for a suitable integer r. It follows, by using Proposition 11 of § 2·6, that $\mathfrak{q} = Q\mathfrak{q} \cap R = Q\mathfrak{p}^r \cap R = \mathfrak{p}^{(r)}$, where $\mathfrak{p}^{(r)}$ is, as usual, the rth symbolic prime power of $\mathfrak{p}$.

THEOREM 9. *Let b be an element of R which is neither a unit nor a zero divisor and let $\mathfrak{p}_1, \mathfrak{p}_2, \ldots, \mathfrak{p}_s$ be the prime ideals which belong to (b). Then $\mathfrak{p}_1, \mathfrak{p}_2, \ldots, \mathfrak{p}_s$ are just the relevant prime ideals which contain b, and, moreover, they are all of them minimal prime ideals of (b). The ideal (b) has only one normal decomposition, and this, with the usual notation for symbolic prime powers, is of the form*

$$(b) = \mathfrak{p}_1^{(r_1)} \cap \mathfrak{p}_2^{(r_2)} \cap \ldots \cap \mathfrak{p}_s^{(r_s)}.$$

Proof. By Theorem 7 and the hypothesis that b is not a zero divisor, it follows that each of $\mathfrak{p}_1, \mathfrak{p}_2, \ldots, \mathfrak{p}_s$ is a relevant prime ideal. Conversely, if $\mathfrak{p}$ is a relevant prime ideal which contains b, then, as has already been observed, $\mathfrak{p}$ is a minimal prime ideal of (b) and therefore $\mathfrak{p}$ occurs among the $\mathfrak{p}_i$. These remarks establish the first two assertions of the theorem, and from the second of the two assertions the uniqueness of the normal decomposition follows immediately. Proposition 8 now shows that the normal decomposition of (b) must be of the form stated.

Let b be an element of R which is not a zero divisor and let $\mathfrak{p}$ be a relevant prime ideal. We now define $\operatorname{ord}_{\mathfrak{p}} b$ as follows: *if $(b) \subseteq \mathfrak{p}$ and if $\mathfrak{p}^{(r)}$ is the then uniquely determined $\mathfrak{p}$-primary component of (b) we put* $\operatorname{ord}_{\mathfrak{p}} b = r$; *if $(b) \nsubseteq \mathfrak{p}$ we put* $\operatorname{ord}_{\mathfrak{p}} b = 0$. It is convenient to define $\mathfrak{p}^{(0)}$ as being the whole ring R.

LEMMA 5. *Let b and c be elements of R which are not zero divisors, let $\mathfrak{p}$ be a relevant prime ideal of R, and let $r \geqslant 0$ be an integer, then*

(1) $\text{ord}_{\mathfrak{p}}\, bc = \text{ord}_{\mathfrak{p}}\, b + \text{ord}_{\mathfrak{p}}\, c$;

(2) *if $b+c$ is not a zero divisor then*

$$\text{ord}_{\mathfrak{p}}\,(b+c) \geqslant \min\,(\text{ord}_{\mathfrak{p}}\, b, \text{ord}_{\mathfrak{p}}\, c)$$

and, further, if $\text{ord}_{\mathfrak{p}}\, b \neq \text{ord}_{\mathfrak{p}}\, c$ *then*

$$\text{ord}_{\mathfrak{p}}\,(b+c) = \min\,(\text{ord}_{\mathfrak{p}}\, b, \text{ord}_{\mathfrak{p}}\, c);$$

(3) *we have* $\text{ord}_{\mathfrak{p}}\, b \geqslant r$ *if, and only if,* $b \in \mathfrak{p}^{(r)}$.

Proof. Let $\text{ord}_{\mathfrak{p}}\, b = \mu$, let $\text{ord}_{\mathfrak{p}}\, c = \nu$, and let Q be the ring of quotients of R with respect to $\mathfrak{p}$. Then $\mathfrak{p}^{(\mu)} = Q(b) \cap R$, which shows that $Q(b) = \mathfrak{m}^{\mu}$; similarly, $Q(c) = \mathfrak{m}^{\nu}$. By Proposition 13 of §2·7, $Q(bc) = Q(b)Q(c) = \mathfrak{m}^{\mu+\nu}$, and this shows that $\text{ord}_{\mathfrak{p}}\,(bc) = \text{ord}_{\mathfrak{p}}\, b + \text{ord}_{\mathfrak{p}}\, c$, as is asserted in (1). We shall now establish (3). If $\mu \geqslant r$ then (trivially) $b \in \mathfrak{p}^{(\mu)} \subseteq \mathfrak{p}^{(r)}$; on the other hand, if $b \in \mathfrak{p}^{(r)}$, then $\mathfrak{m}^{\mu} = Q(b) \subseteq Q\mathfrak{p}^{(r)} = \mathfrak{m}^{r}$, and this shows that $\mu \geqslant r$. Finally, suppose that $\mu \leqslant \nu$ and that $b+c$ is not a zero divisoı, then $\mathfrak{p}^{(\nu)} \subseteq \mathfrak{p}^{(\mu)}$ and $b+c \in \mathfrak{p}^{(\mu)} + \mathfrak{p}^{(\nu)} = \mathfrak{p}^{(\mu)}$, consequently, by (3), $\text{ord}_{\mathfrak{p}}\,(b+c) \geqslant \mu = \min\,(\mu, \nu)$. If, however, $\mu < \nu$ then, by (3), $b \notin \mathfrak{p}^{(\mu+1)}$, while $c \in \mathfrak{p}^{(\nu)} \subseteq \mathfrak{p}^{(\mu+1)} \subseteq \mathfrak{p}^{(\mu)}$. It follows that $b+c$ belongs to $\mathfrak{p}^{(\mu)}$ but not to $\mathfrak{p}^{(\mu+1)}$, and this shows that $\text{ord}_{\mathfrak{p}}\,(b+c)$ is precisely μ. The proof is now complete.

These results will be put into a new form which resembles the calculus of functions and their divisors in algebraic geometry. For notational convenience, we first index the relevant prime ideals by putting them into 1-1 correspondence with a set I of symbols, and then if $i \in I$ we shall denote the corresponding relevant prime ideal by $\mathfrak{p}_i$. By a *divisor* $\sum_{i \in I} s_i \mathfrak{p}_i$ (or, more simply, $\Sigma s_i \mathfrak{p}_i$), we mean a formal sum in which the 'coefficients' s_i are integers (positive, negative or zero), and in which at most a finite number of coefficients are different from zero. If $\Sigma s_i \mathfrak{p}_i$ and $\Sigma t_i \mathfrak{p}_i$ are divisors, then $\Sigma (s_i + t_i)\, \mathfrak{p}_i$ and $\Sigma (s_i - t_i)\, \mathfrak{p}_i$ are also divisors, and so we can define addition and subtraction of divisors by

$$\Sigma s_i \mathfrak{p}_i + \Sigma t_i \mathfrak{p}_i = \Sigma (s_i + t_i)\, \mathfrak{p}_i,$$

$$\Sigma s_i \mathfrak{p}_i - \Sigma t_i \mathfrak{p}_i = \Sigma (s_i - t_i)\, \mathfrak{p}_i.$$

The divisor in which all the coefficients are zero is called the *null divisor*, and a divisor in which all the coefficients are non-negative is called an *integral divisor*.

If an element b of R is not a zero divisor we shall write $v_i(b) = \text{ord}_{\mathfrak{p}_i} b$. This definition can be extended as follows. Let α belong to the ring $\mathfrak{R}$ of quotients and assume that α is not a zero divisor, then $\alpha = b/c$, where b and c are both in R and neither of them is a zero divisor. Put $v_i(\alpha) = v_i(b) - v_i(c)$, then $v_i(\alpha)$ is (by (1) of Lemma 5) independent of the way in which α is written as a quotient, and v_i retains its original meaning on R. It is now possible to extend (1) and (2) of Lemma 5; i.e. *if α and β are in $\mathfrak{R}$ and neither is a zero divisor then $v_i(\alpha\beta) = v_i(\alpha) + v_i(\beta)$ and (assuming that $\alpha+\beta$ is not a zero divisor) $v_i(\alpha+\beta) \geqslant \min[v_i(\alpha), v_i(\beta)]$ with equality if $v_i(\alpha) \neq v_i(\beta)$.* By the definition of $\text{ord}_{\mathfrak{p}_i} b$ we see that $v_i(b)$ is different from zero if, and only if, $\mathfrak{p}_i$ belongs to (b), hence $v_i(b)$, and similarly $v_i(c)$, is different from zero for only a finite number of different values of i. But $v_i(\alpha) = v_i(b) - v_i(c)$, consequently $\Sigma v_i(\alpha)\,\mathfrak{p}_i$ is a divisor; we call $\Sigma v_i(\alpha)\,\mathfrak{p}_i$ *the divisor of* α and we shall write $\alpha \sim \Sigma v_i(\alpha)\,\mathfrak{p}_i$.

THEOREM 10. *Let $\alpha \sim \Sigma s_i \mathfrak{p}_i$, and let $\beta \sim \Sigma t_i \mathfrak{p}_i$, then*

$$\alpha\beta \sim \Sigma s_i \mathfrak{p}_i + \Sigma t_i \mathfrak{p}_i \quad \textit{and} \quad \alpha/\beta \sim \Sigma s_i \mathfrak{p}_i - \Sigma t_i \mathfrak{p}_i.$$

Further, α belongs to R if, and only if, its divisor is integral.

Proof. Since $v_i(\alpha\beta) = v_i(\alpha) + v_i(\beta) = s_i + t_i$, the first assertion follows from the definition of the sum of two divisors. It is clear that $v_i(\beta^{-1}) = -v_i(\beta)$, consequently $v_i(\alpha/\beta) = v_i(\alpha) - v_i(\beta) = s_i - t_i$, and the second assertion follows. If $\alpha \in R$ then certainly $v_i(\alpha) = s_i \geqslant 0$ for all i and therefore the divisor of α is integral. Finally, let us assume that $v_i(\alpha) \geqslant 0$ for all i. Choose b and c in R so that neither is a zero divisor and so that $\alpha = b/c$, then if $v_i(b) = p_i$ and $v_i(c) = q_i$ we have $p_i \geqslant q_i$ for all i. By (3) of Lemma 5, it follows that $b \in \mathfrak{p}_i^{(q_i)}$ for all i, hence, since

$$(c) = \bigcap_{i \in I,\, q_i > 0} \mathfrak{p}_i^{(q_i)},$$

$b \in (c)$. This shows that $\alpha = b/c$ belongs to R.

COROLLARY. *Suppose that neither α nor β is a zero divisor in $\mathfrak{R}$, then α and β have the same divisor if, and only if, α/β is a unit in R.*

Proof. Let $\gamma=\alpha/\beta$, then α and β have the same divisor if, and only if, the divisor of γ is the null divisor. Now the divisor of γ will be null if, and only if, both γ and γ^{-1} have integral divisors, and this (by the theorem) will be the case if, and only if, γ and γ^{-1} are both in R.

4·10. Another theorem on divisors. The divisors which belong to elements form only a part of the set of all divisors, and the question arises as to how far the divisor of an element may be prescribed. Before we attempt to discuss this question we shall impose an additional condition on R, namely, we shall assume that R is an integral domain. For the reader's convenience we recapitulate our hypotheses. They amount to this: *in what follows, R is a Noetherian integral domain which is integrally closed in its quotient field $\mathfrak{R}$.*

LEMMA 6. *Let I_0 be a finite subset of I and let i belong to I but not to I_0, then we can find $b \in R$ $(b \neq 0)$ such that $v_i(b)=1$ and $v_j(b)=0$ for all $j \in I_0$.*

Proof. Choose $x \in \mathfrak{p}_i$ so that $x \notin \mathfrak{p}_i^{(2)}$ and then decompose I_0 into two subsets I_1 and I_2 where $x \in \mathfrak{p}_j$ if $j \in I_1$ and $x \notin \mathfrak{p}_k$ if $k \in I_2$. Since all relevant prime ideals have rank one, it is not possible for one of them to contain another, consequently $\mathfrak{p}_j$ cannot contain

$$\mathfrak{p}_i^{(2)} \cap \left(\bigcap_{k \in I_2} \mathfrak{p}_k\right)$$

if $j \in I_1$. We can therefore (Proposition 6, § 1·5) choose y so that

$$y \in \mathfrak{p}_i^{(2)} \cap \left(\bigcap_{k \in I_2} \mathfrak{p}_k\right)$$

and also $y \notin \mathfrak{p}_j$ if $j \in I_1$. Put $b=x+y$. Then $b \in \mathfrak{p}_i+\mathfrak{p}_i^{(2)}=\mathfrak{p}_i$, but, since $x \notin \mathfrak{p}_i^{(2)}$ and $y \in \mathfrak{p}_i^{(2)}$, $b \notin \mathfrak{p}_i^{(2)}$. This proves, incidentally, that $b \neq 0$ and therefore, since R is an integral domain, b is not a zero divisor. By (3) of Lemma 5, $v_i(b)=1$. Again, if $j \in I_1$, then $x \in \mathfrak{p}_j$ and $y \notin \mathfrak{p}_j$ and therefore $b=x+y \notin \mathfrak{p}_j$, consequently $v_j(b)=0$. Finally, if $k \in I_2$ then $x \notin \mathfrak{p}_k$ and $y \in \mathfrak{p}_k$, so $b \notin \mathfrak{p}_k$ and $v_k(b)=0$.

THEOREM 11. *Let I_0 be a finite subset of I and suppose that an integer s_i, which may be positive, negative or zero, is given for each*

$i \in I_0$. Then there exists an element α ($\alpha \neq 0$) in the quotient field of R such that $v_i(\alpha) = s_i$ for all $i \in I_0$ and $v_j(\alpha) \geqslant 0$ for all $j \in I - I_0$.

Proof. For each $i \in I_0$ we can choose, by Lemma 6, an element $b_i \in R$ such that $v_i(b_i) = 1$ and such that $v_{i'}(b_i) = 0$ if $i' \in I_0$ and $i' \neq i$. Put $\beta = \prod_{i \in I_0} b_i^{s_i}$, then $v_i(\beta) = s_i$ for all $i \in I_0$. Now choose a finite subset I_1 of I such that I_0 and I_1 have no common element and $v_k(\beta) = 0$ if k is neither in I_0 nor in I_1. By Lemma 6, for each $j \in I_1$ we can choose $c_j \in R$ so that $v_j(c_j) = 1$ and $v_i(c_j) = 0$ if $i \in I_0$. Put $d = \prod_{j \in I_1} c_j$ then $d \in R$. We consider the element βd^N where N is a positive integer. If $i \in I_0$ then

$$v_i(\beta d^N) = v_i(\beta) + N v_i(d) = v_i(\beta) = s_i;$$

if $j \in I_1$ then

$$v_j(\beta d^N) = v_j(\beta) + N v_j(d) \geqslant v_j(\beta) + N v_j(c_j),$$

consequently
$$v_j(\beta d^N) \geqslant v_j(\beta) + N;$$

if k is neither in I_0 nor in I_1 then

$$v_k(\beta d^N) \geqslant v_k(\beta) = 0.$$

Thus $\alpha = \beta d^N$ has all the required properties if N is sufficiently large.

4·11. Continuation of § 4·3. The reader will recall that the proofs of Propositions 2 and 5 were postponed in order to avoid having a large break in the main discussion. These proofs will now be given. Let R be a Noetherian ring, z a single indeterminate, $\mathfrak{p}$ a prime ideal in R and $\mathfrak{q}$ a $\mathfrak{p}$-primary ideal. Put $R^* = R[z]$, $\mathfrak{p}^* = R^*\mathfrak{p}$ and $\mathfrak{q}^* = R^*\mathfrak{q}$. We shall prove that $\mathfrak{p}^*$ is prime, that $\mathfrak{q}^*$ is $\mathfrak{p}^*$-primary, and that rank $\mathfrak{p}^*$ = rank $\mathfrak{p}$. In other words, we shall prove Propositions 2 and 5 for the case in which R^* is obtained from R by adjoining just one indeterminate; this will be sufficient, because, in order to obtain the full result, we have only to adjoin a number of indeterminates in succession. In the proofs which follow, $\phi(z)$ and $\psi(z)$ will denote polynomials with coefficients in R.

$\mathfrak{p}^*$ *is prime*. Assume that $\phi(z)\psi(z) \in \mathfrak{p}^*$, that $\phi(z) \notin \mathfrak{p}^*$ and that $\psi(z) \notin \mathfrak{p}^*$. We shall obtain a contradiction. Let

$$\phi(z) = a_0 + a_1 z + \ldots + a_s z^s$$

and let $$\psi(z)=b_0+b_1z+\ldots+b_tz^t,$$

then not every a_i is in $\mathfrak{p}$ and not every b_j is in $\mathfrak{p}$. Let a_l be the first of $a_0, a_1, \ldots, a_s$ which is not in $\mathfrak{p}$ and let b_m be the first of $b_0, b_1, \ldots, b_t$ which is not in $\mathfrak{p}$. Put

$$c=a_{l+m}b_0+\ldots+a_lb_m+\ldots+a_0b_{l+m},$$

then, since $a_lb_m \notin \mathfrak{p}$ while every other term in the sum belongs to $\mathfrak{p}$, $c\notin\mathfrak{p}$. But c is the coefficient of z^{l+m} in $\phi(z)\psi(z)$, consequently, by Lemma 2, $\phi(z)\psi(z)\notin\mathfrak{p}^*$. This is the required contradiction.

$\mathfrak{q}^*$ *is* $\mathfrak{p}^*$-*primary*. Let $\mathfrak{p}'$ be any prime ideal belonging to $\mathfrak{q}^*$, then $\mathfrak{p}'\cap R \supseteq \mathfrak{q}^*\cap R=\mathfrak{q}$ (Lemma 2, Corollary 1). But $\mathfrak{q}$ is $\mathfrak{p}$-primary and $\mathfrak{p}'\cap R$ is prime, accordingly $\mathfrak{p}\subseteq\mathfrak{p}'\cap R$ and therefore $\mathfrak{p}^*=R^*\mathfrak{p}\subseteq\mathfrak{p}'$. We shall show that $\mathfrak{p}'=\mathfrak{p}^*$. Suppose that $\phi(z)=a_hz^h+a_{h+1}z^{h+1}+\ldots+a_kz^k$ belongs to $\mathfrak{p}'$, then, by Theorem 6 of §1·9, $\mathfrak{q}^*:(\phi)\neq\mathfrak{q}^*$, and therefore we can find $\psi(z)=b_rz^r+b_{r+1}z^{r+1}+\ldots+b_sz^s$ such that (i) $\phi(z)\psi(z)\in\mathfrak{q}^*$ and (ii) $\psi(z)\notin\mathfrak{q}^*$. Let us observe that if $b_r, b_{r+1}, \ldots, b_{r+j}$ are all in $\mathfrak{q}$ then we can omit the terms $b_rz^r, b_{r+1}z^{r+1}, \ldots, b_{r+j}z^{r+j}$ from $\psi(z)$ without disturbing the two essential properties; consequently, since not all the coefficients of $\psi(z)$ can be in $\mathfrak{q}$, it is possible to arrange that $b_r\not\equiv 0(\mathfrak{q})$. Now it follows from $\phi(z)\psi(z)\in\mathfrak{q}^*$ that all the coefficients of $\phi(z)\psi(z)$ are in $\mathfrak{q}$ (see Lemma 2) and therefore, in particular, $a_hb_r\in\mathfrak{q}$. But $\mathfrak{q}$ is $\mathfrak{p}$-primary and $b_r\notin\mathfrak{q}$, consequently $a_h\in\mathfrak{p}$. Further, $a_{h+1}z^{h+1}+\ldots+a_kz^k=\phi(z)-a_hz^h\in\mathfrak{p}'+\mathfrak{p}^*=\mathfrak{p}'$, and so we can show, by a repetition of our arguments, that $a_{h+1}\in\mathfrak{p}$. Proceeding in this way, we find that all the coefficients of $\phi(z)$ are in $\mathfrak{p}$ and hence that $\phi(z)\in\mathfrak{p}^*$. But $\phi(z)$ was any element of $\mathfrak{p}'$ and we already know that $\mathfrak{p}^*\subseteq\mathfrak{p}'$, consequently $\mathfrak{p}^*=\mathfrak{p}'$. Thus there is only one prime ideal belonging to $\mathfrak{q}^*$, namely, $\mathfrak{p}^*$, and therefore $\mathfrak{q}^*$ must be $\mathfrak{p}^*$-primary.

rank $\mathfrak{p}$ = *rank* $\mathfrak{p}^*$. We have now established Proposition 2 and therefore, in proving that rank $\mathfrak{p}$ is equal to rank $\mathfrak{p}^*$, we may legitimately use any result of §4·3 which occurs before Proposition 5. For this reason, the following result, which is obtained by combining Proposition 3 with Corollary 1 of Lemma 2, may be used for our present purposes: *if* $\mathfrak{p}$ *is a minimal prime ideal of an ideal* $\mathfrak{a}$, *then* $\mathfrak{p}^*=R^*\mathfrak{p}$ *is a minimal prime ideal of* $R^*\mathfrak{a}$. Let rank $\mathfrak{p} = r$ and let rank $\mathfrak{p}^* = s$. If $r = 0$ then $\mathfrak{p}$ is a minimal prime ideal of the zero ideal, consequently $\mathfrak{p}^*$ is a minimal prime ideal of the zero ideal (in R^*) and therefore s is also zero. We shall now suppose that $r \geqslant 1$. By the definition of rank, there exist prime ideals $\mathfrak{p}_1, \mathfrak{p}_2, \ldots, \mathfrak{p}_r$ such that $\mathfrak{p} \supset \mathfrak{p}_1 \supset \mathfrak{p}_2 \supset \ldots \supset \mathfrak{p}_r$. Put $\mathfrak{p}_i^* = R^*\mathfrak{p}_i$, then $\mathfrak{p}_i^*$ is prime and $\mathfrak{p}^* \supset \mathfrak{p}_1^* \supset \mathfrak{p}_2^* \supset \ldots \supset \mathfrak{p}_r^*$ which shows that $s \geqslant r$. Now, by Theorem 8 of §3·5, we can find r elements $a_1, a_2, \ldots, a_r$ of $\mathfrak{p}$ such that $(a_1, a_2, \ldots, a_r)$ is of rank r, and then, because rank $\mathfrak{p} = r$, $\mathfrak{p}$ must be a minimal prime ideal of $(a_1, a_2, \ldots, a_r)$. This shows that $\mathfrak{p}^*$ is a minimal prime ideal of

$$R^*a_1 + R^*a_2 + \ldots + R^*a_r,$$

consequently, by Theorem 7 of §3·5, $s \leqslant r$. The proof is now complete.

CHAPTER V

THE ANALYTIC THEORY OF LOCAL RINGS

5·1. Convergence. It will be shown in this chapter that it is possible to build up a far-reaching theory of limits in any given local ring Q. This theory will enable us to associate with Q a certain other local ring $\bar{Q}$, which is called the *completion* of Q. The way in which $\bar{Q}$ is obtained from Q is very similar to that in which the real numbers are obtained from the rational numbers; accordingly, it is natural to describe as *analytic*, results and methods which make use of $\bar{Q}$.

To define limits and convergence in Q, we proceed as follows. Let $b_1, b_2, b_3, \ldots$ be an infinite sequence of elements of the local ring, and let b also belong to Q. We shall say that *the sequence* (b_n) *converges to* b, and that b_n *tends to* b *as* n *tends to infinity*, if the following condition is satisfied: *Given any integer* $s \geqslant 0$, *we can always find an integer* $n_0 = n_0(s)$ *such that* $b - b_n \in \mathfrak{m}^s$ *for all* $n > n_0$. Just as in classical analysis, instead of writing the rather lengthy phrase 'b_n tends to b as n tends to infinity' we shall often write $b_n \to b$ as $n \to \infty$, or, still more simply, $b_n \to b$. Other well-known conventions and notations of analysis will be adapted in a similar way, and, since the intention will usually be quite clear, explanations will often be omitted; for example, we shall use $\lim b_n = b$ as an alternative for $b_n \to b$. There is one point, however, that needs further attention before we proceed, namely, we must show that if $b_n \to b$ and also $b_n \to b'$ then $b = b'$. This can be proved as follows. Let $s \geqslant 0$ be a given integer, then we can find integers n_1 and n_2 such that $b - b_n \in \mathfrak{m}^s$ if $n > n_1$ and $b' - b_n \in \mathfrak{m}^s$ if $n > n_2$. Choose $n > \max(n_1, n_2)$ then $b - b' = (b - b_n) + (b_n - b') \in \mathfrak{m}^s$, hence since s was arbitrary, it follows by (4·1·1) that $b = b'$.

Proposition 1 asserts that addition, subtraction and multiplication are 'continuous' operations.

PROPOSITION 1. *Suppose that* $a_n \to a$ *and that* $b_n \to b$, *then* $a_n + b_n \to a + b$, $a_n - b_n \to a - b$, *and* $a_n b_n \to ab$.

The proofs of these assertions are almost trivial. As an example we shall show that $a_n b_n \to ab$. Let $s \geqslant 0$ be given, then we can find an integer n_0 such that $a - a_n \in \mathfrak{m}^s$ and $b - b_n \in \mathfrak{m}^s$ for all $n > n_0$. If now $n > n_0$ we have $ab - a_n b_n = (a - a_n) b + a_n(b - b_n) \in \mathfrak{m}^s$, which shows that $a_n b_n \to ab$.

PROPOSITION 2. *Suppose that (a_n) is a sequence of elements all of which belong to an ideal $\mathfrak{a}$, and suppose also that $a_n \to x$, then $x \in \mathfrak{a}$.*

Proof. By (4·2·1) it is enough to show that if s is a non-negative integer, then $x \in \mathfrak{a} + \mathfrak{m}^s$. But if we choose n so large that $x - a_n \in \mathfrak{m}^s$ we have $x = a_n + (x - a_n) \in \mathfrak{a} + \mathfrak{m}^s$ as required.

5·2. Complete rings. Let (a_n) be a sequence of elements in a local ring Q.

DEFINITION. *The sequence (a_n) is said to be a 'Cauchy sequence' if, given any integer $s \geqslant 0$, we can always find an integer $n_0 = n_0(s)$ such that $a_n - a_m \in \mathfrak{m}^s$ whenever $n > m > n_0$.*

The Cauchy sequences, which occur in connexion with local rings, are in some ways easier to handle than those which occur in ordinary analysis, because we have

LEMMA 1. *The sequence (a_n) is a Cauchy sequence if, and only if, $a_n - a_{n-1} \to 0$ as $n \to \infty$.*

Proof. First, suppose that (a_n) is a Cauchy sequence. Let $s \geqslant 0$ be given and choose n_0 so that $a_n - a_m \in \mathfrak{m}^s$ if $n > m > n_0$; then, if $n > n_0 + 1$ we have $a_n - a_{n-1} \in \mathfrak{m}^s$, consequently $a_n - a_{n-1} \to 0$. Next, suppose that $a_n - a_{n-1} \to 0$. If $s \geqslant 0$ is given we can choose n_0 so that $a_n - a_{n-1} \in \mathfrak{m}^s$ for all $n > n_0$. Assume that $n > m > n_0$, then $a_{m+1} - a_m, a_{m+2} - a_{m+1}, \ldots, a_n - a_{n-1}$ are all in $\mathfrak{m}^s$, consequently

$$a_n - a_m = (a_{m+1} - a_m) + (a_{m+2} - a_{m+1}) + \ldots + (a_n - a_{n-1})$$

is in $\mathfrak{m}^s$. This proves that (a_n) is a Cauchy sequence.

Suppose that $a_n \to a$ as $n \to \infty$, then $a_n - a_{n-1} \to a - a = 0$; this, when combined with Lemma 1, shows that *if a sequence (a_n) has a limit in Q then (a_n) is a Cauchy sequence.* This, at once, suggests the

DEFINITION. *A local ring Q is said to be 'complete' if every Cauchy sequence composed of elements of Q has a limit in Q.*

PROPOSITION 3. *If Q is a complete local ring then the series $\sum_{1}^{\infty} a_n$ converges if, and only if, a_n tends to zero as n tends to infinity.*

Proof. Let $A_n = a_1 + a_2 + \ldots + a_n$, then A_n tends to a limit if, and only if, (A_n) is a Cauchy sequence. But, by Lemma 1, (A_n) is a Cauchy sequence if, and only if, $A_n - A_{n-1} = a_n \to 0$ as $n \to \infty$.

PROPOSITION 4. *Let $\mathfrak{a}$ be a proper ideal in the local ring Q, let $Q' = Q/\mathfrak{a}$, and let σ be the natural homomorphism of Q upon Q'. Then if $a_n \to a$ in Q we have $\sigma(a_n) \to \sigma(a)$ in Q'. Also if Q is complete so is Q'.*

Proof. Let $s \geqslant 0$ be given. Choose n_0 so that $a - a_n \in \mathfrak{m}^s$ for all $n > n_0$, then $\sigma(a) - \sigma(a_n) \in \sigma(\mathfrak{m}^s) = \mathfrak{m}'^s$ for all $n > n_0$, which shows that $\sigma(a_n) \to \sigma(a)$. Now assume that Q is complete and let (a_n') be a Cauchy sequence in Q'. Put $b_n' = a_n' - a_{n-1}'$ for $n \geqslant 2$ and put $b_1' = a_1'$. By Lemma 1, $b_n' \to 0$, consequently we can find integers $s_1, s_2, s_3, \ldots$, such that $s_n \to \infty$ as $n \to \infty$ and $b_n' \in \mathfrak{m}'^{s_n}$ for all n. Since $\sigma(\mathfrak{m}^s) = \mathfrak{m}'^s$, we can choose $b_n \in \mathfrak{m}^{s_n}$ such that $\sigma(b_n) = b_n'$, and then, by construction $b_n \to 0$. Now Q is complete, consequently $a_n = b_1 + b_2 + \ldots + b_n$ tends to a limit a (say) as $n \to \infty$ (Proposition 3), and therefore, by the first part, $\sigma(a_n) = b_1' + b_2' + \ldots + b_n' = a_n'$ tends to $\sigma(a)$. This completes the proof.

We come now to an important but more difficult result.

THEOREM 1. *Let Q be a complete local ring and let $\mathfrak{b}_n$ $(n = 1, 2, \ldots)$ be a decreasing sequence of ideals $(\mathfrak{b}_n \supseteq \mathfrak{b}_{n+1})$ such that $\bigcap_{n=1}^{\infty} \mathfrak{b}_n = (0)$. Then given an integer $h \geqslant 0$, however large, we can always find an integer $k \geqslant 0$ such that $\mathfrak{b}_k \subseteq \mathfrak{m}^h$.*

Proof. We may suppose that all the $\mathfrak{b}_n$ are proper ideals, because there are, at most, only a finite number of non-proper $\mathfrak{b}_n$, and these may be left out of the sequence. With this proviso,

$$\mathfrak{b}_1 + \mathfrak{m}^s \supseteq \mathfrak{b}_2 + \mathfrak{m}^s \supseteq \mathfrak{b}_3 + \mathfrak{m}^s \supseteq \ldots$$

is a decreasing sequence of $\mathfrak{m}$-primary ideals all of which contain $\mathfrak{m}^s$, consequently the number of different terms in the sequence is finite; in fact, this number does not exceed the length of $\mathfrak{m}^s$. Choose an integer n_s so that

$$\mathfrak{b}_n + \mathfrak{m}^s = \mathfrak{b}_{n_s} + \mathfrak{m}^s \quad \text{for all } n \geqslant n_s. \tag{A}$$

We now have a sequence $n_1, n_2, n_3, \ldots$ of integers, and it is clear that we can arrange that $n_i < n_{i+1}$ for all i. In order to simplify our notations, we put $\mathfrak{b}_{n_s} = \mathfrak{c}_s$, then the $\mathfrak{c}_s$ form a decreasing sequence and they have only the zero element in common. Further, by (A), we have, on putting $n = n_{s+1}$,

$$\mathfrak{c}_{s+1} + \mathfrak{m}^s = \mathfrak{c}_s + \mathfrak{m}^s. \tag{B}$$

We shall prove that for each r we have $\mathfrak{c}_r \subseteq \mathfrak{m}^r$, and this will establish the theorem. We therefore take a fixed value of r and suppose that $x \in \mathfrak{c}_r$; it remains for us to show that $x \in \mathfrak{m}^r$. By (B), $x \in \mathfrak{c}_r \subseteq \mathfrak{c}_{r+1} + \mathfrak{m}^r$, hence $x = y_{r+1} + a_r$, where $y_{r+1} \in \mathfrak{c}_{r+1}$, $a_r \in \mathfrak{m}^r$. Again, $y_{r+1} \in \mathfrak{c}_{r+1} \subseteq \mathfrak{c}_{r+2} + \mathfrak{m}^{r+1}$, consequently $y_{r+1} = y_{r+2} + a_{r+1}$, where $y_{r+2} \in \mathfrak{c}_{r+2}$, $a_{r+1} \in \mathfrak{m}^{r+1}$. We now have $x = y_{r+2} + a_r + a_{r+1}$. Next write y_{r+2} in the form $y_{r+2} = y_{r+3} + a_{r+2}$, where $y_{r+3} \in \mathfrak{c}_{r+3}$, $a_{r+2} \in \mathfrak{m}^{r+2}$. In this manner we obtain sequences $y_{r+1}, y_{r+2}, y_{r+3}, \ldots$ and $a_r, a_{r+1}, a_{r+2}, \ldots$ where $y_{r+j} \in \mathfrak{c}_{r+j}$, $a_{r+i} \in \mathfrak{m}^{r+i}$, and where $x = y_{r+n} + a_r + a_{r+1} + \ldots + a_{r+n-1}$ for all n. Since $a_{r+n} \to 0$ as $n \to \infty$, it follows, by Proposition 3, that $a_r + a_{r+1} + \ldots + a_{r+n-1}$ tends to a limit α (say) as $n \to \infty$; consequently $y_{r+n} \to u$, as $n \to \infty$, where $x = u + \alpha$. But $a_r + a_{r+1} + \ldots + a_{r+n-1} \in \mathfrak{m}^r$, accordingly $\alpha \in \mathfrak{m}^r$ by Proposition 2. Further, by the same proposition, since y_{r+p}, y_{r+p+1}, y_{r+p+2}, etc. are all in $\mathfrak{c}_{r+p}$, it follows that $u \in \mathfrak{c}_{r+p}$ for every positive integral value of p. This, however, means that $u = 0$, consequently $x = \alpha \in \mathfrak{m}^r$ as was required.

5·3. Concordant rings. If a local ring Q' is an extension ring of a local ring Q it becomes necessary to know whether or not limit operations in Q are compatible with those in Q'.

DEFINITION. *If a sequence of elements of Q is a Cauchy sequence in Q when, and only when, it is a Cauchy sequence in Q', then we say that Q' is a 'concordant extension' of Q.*

We now give two alternative forms of the 'condition of concordancy'.

LEMMA 2. *Let the local ring Q' be an extension ring of the local ring Q, then each of the following two conditions is necessary and sufficient for Q' to be a concordant extension of Q.*

(1) *A sequence of elements in Q tends to zero in Q if, and only if, it tends to zero in Q'.*

(2) *Given any two integers $h \geqslant 0$ and $k \geqslant 0$, we can find integers $r \geqslant 0$ and $s \geqslant 0$ such that $\mathfrak{m}^r \subseteq \mathfrak{m}'^h \cap Q$ and $\mathfrak{m}'^s \cap Q \subseteq \mathfrak{m}^k$.*

Proof. First assume that Q' is a concordant extension of Q, let (a_n) be a sequence of elements of Q, and put $A_n = a_1 + a_2 + \ldots + a_n$. If $a_n \to 0$ in Q, then A_n is a Cauchy sequence in Q (Lemma 1), and therefore, by assumption, A_n is a Cauchy sequence in Q'. It now follows, by Lemma 1, that $a_n = A_n - A_{n-1}$ tends to zero in Q'. In a similar manner, we can show that if $a_n \to 0$ in Q' then $a_n \to 0$ in Q. Thus concordancy implies (1).

Next, assume that Q and Q' have the property stated in (1) and let integers $h \geqslant 0$ and $k \geqslant 0$ be given. Then, for some r, $\mathfrak{m}^r \subseteq \mathfrak{m}'^h \cap Q$; otherwise we could choose, for each r, an element $a_r \in \mathfrak{m}^r$ such that $a_r \notin \mathfrak{m}'^h$. The sequence (a_r) would then be such that $a_r \to 0$ in Q but not in Q', and this would be a contradiction. A similar reasoning shows that $\mathfrak{m}'^s \cap Q \subseteq \mathfrak{m}^k$ for some s. Thus (1) implies (2).

Finally, assume that (2) is true, and let (a_n) be a sequence of elements of Q. Suppose, for example, that (a_n) is a Cauchy sequence in Q' and let an integer $k \geqslant 0$ be given. Choose s so that $\mathfrak{m}'^s \cap Q \subseteq \mathfrak{m}^k$, and then choose n_0 so that $a_n - a_m \in \mathfrak{m}'^s$ if $n > m > n_0$. If now $n > m > n_0$, we have $a_n - a_m \in \mathfrak{m}'^s \cap Q \subseteq \mathfrak{m}^k$, which shows that (a_n) is a Cauchy sequence in Q. Similarly, if (a_n) is a Cauchy sequence in Q then it is also a Cauchy sequence in Q'. Thus (2) implies concordancy. This completes the proof.

If Q' is a concordant extension of Q, then, by (2) of Lemma 2, we can find an integer r such that $\mathfrak{m}^r \subseteq \mathfrak{m}' \cap Q$. Since $\mathfrak{m}' \cap Q$ is a prime ideal (§2·4, Proposition 8), it follows that $\mathfrak{m} \subseteq \mathfrak{m}' \cap Q$, and therefore, since $\mathfrak{m}$ is maximal and $1 \notin \mathfrak{m}'$, $\mathfrak{m} = \mathfrak{m}' \cap Q$. Thus $\mathfrak{m} = \mathfrak{m}' \cap Q$ *is a necessary condition for Q' to be a concordant extension of Q.* It will now be proved that, if Q is complete, this condition is also sufficient.

THEOREM 2. *Let Q be a complete local ring and let Q' be a local ring which is an extension ring of Q. Then Q' is a concordant extension if, and only if, $\mathfrak{m} = \mathfrak{m}' \cap Q$.*

Proof. We already know that the condition $\mathfrak{m} = \mathfrak{m}' \cap Q$ is necessary for concordancy. Assume then that $\mathfrak{m} = \mathfrak{m}' \cap Q$ and let

$h \geqslant 0$, $k \geqslant 0$ be given integers. Since $\mathfrak{m} = \mathfrak{m}' \cap Q$, we have $\mathfrak{m}^h \subseteq \mathfrak{m}'^h \cap Q$, consequently it is enough to show that there is an integer s such that $\mathfrak{m}'^s \cap Q \subseteq \mathfrak{m}^k$ (Lemma 2). But $\mathfrak{m}'^s \cap Q$ is a decreasing sequence of ideals, consequently, by Theorem 1, it is enough to show that the ideals $\mathfrak{m}'^s \cap Q$ have only the zero element in common. This, however, is obvious, because $\bigcap_{s=1}^{\infty} \mathfrak{m}'^s = (0)$.

5·4. Formal power series. If R is a ring and $X_1, X_2, \ldots, X_n$ are indeterminates, then, besides the polynomials in the X_i with coefficients in R, we can also consider power series in the X_i with coefficients in R. Let $(\nu) = (\nu_1, \nu_2, \ldots, \nu_n)$ be an ordered set of n non-negative integers, then a natural notation for a power series is obtained by writing it as $\sum_{(\nu)} a_{(\nu)} X_1^{\nu_1} X_2^{\nu_2} \ldots X_n^{\nu_n}$, where each coefficient $a_{(\nu)}$ is in R. A power series is determined when all the coefficients $a_{(\nu)}$ are known; the $a_{(\nu)}$ may be quite unrestricted (provided that they are in R); and two power series

$$\sum_{(\nu)} a_{(\nu)} X_1^{\nu_1} X_2^{\nu_2} \ldots X_n^{\nu_n} \quad \text{and} \quad \sum_{(\nu)} b_{(\nu)} X_1^{\nu_1} X_2^{\nu_2} \ldots X_n^{\nu_n}$$

are to be regarded as being the same if, and only if, $a_{(\nu)} = b_{(\nu)}$ for all (ν). Let us define addition and multiplication of power series by the obvious rules. The power series now form a ring, which is usually denoted by $R[[X_1, X_2, \ldots, X_n]]$ and which is called *the ring of formal power series in* $X_1, X_2, \ldots, X_n$ *with coefficients in* R. Each element of R can be regarded as a 'constant power series', so that $R[[X_1, X_2, \ldots, X_n]]$ can be considered to be an extension ring of R. If k is an integer and if each of the non-zero terms of the power series ϕ has a degree which is greater than k, then we shall write $\phi = o(k)$; thus $\phi = o(k)$ if, and only if, $\phi \in (X_1, X_2, \ldots, X_n)^{k+1}$.

Rings of formal power series, especially when the coefficients are taken from a field, are of great interest for their own sake. Here they have been introduced in preparation for the proof, which will be given in the next section, that every local ring has a completion.

THEOREM 3. *If R is a Noetherian ring, then $R[[X_1, X_2, \ldots, X_n]]$ is also Noetherian.*

Proof. Put $R_0 = R$ and $R_i = R[[X_1, X_2, \ldots, X_i]]$ $(1 \leqslant i \leqslant n)$, then R_i can be considered as a ring of formal power series in X_i with coefficients in R_{i-1}; $R_i = R_{i-1}[[X_i]]$. This shows that we need only consider power series in a single variable X, for, once the theorem has been proved in this special case, the general result will follow by induction.

Let $\mathfrak{A}$ be an ideal in $R[[X]]$. We form a set $\mathfrak{a}_k$ $(k = 0, 1, 2, \ldots)$ of elements of R, by putting an element $a \in R$ in $\mathfrak{a}_k$ if, and only if, there is a power series $\phi \in \mathfrak{A}$ which is of the form $\phi = aX^k + o(k)$. It is clear that $\mathfrak{a}_k$ is an ideal; further, since $X\phi = aX^{k+1} + o(k+1)$ belongs to $\mathfrak{A}$, we see that $\mathfrak{a}_k \subseteq \mathfrak{a}_{k+1}$. But R is Noetherian, therefore we can choose m so that $\mathfrak{a}_\mu = \mathfrak{a}_m$ for all $\mu \geqslant m$. Let us now select a base for each of the ideals $\mathfrak{a}_0, \mathfrak{a}_1, \ldots, \mathfrak{a}_m$ and, for notational convenience, let us arrange that each base contains the same number s (say) of elements; thus $\mathfrak{a}_i = (a_1^{(i)}, a_2^{(i)}, \ldots, a_s^{(i)})$. By construction, if $0 \leqslant i \leqslant m$ and $1 \leqslant j \leqslant s$, there is a power series $\phi_j^{(i)} \in \mathfrak{A}$ such that $\phi_j^{(i)} = a_j^{(i)} X^i + o(i)$. We shall show that the $\phi_j^{(i)}$ are a base of $\mathfrak{A}$. Let $\phi \in \mathfrak{A}$. If a_0 is the 'constant term' of ϕ then $a_0 \in \mathfrak{a}_0$, consequently we can find elements $r_j^{(0)}$ $(1 \leqslant j \leqslant s)$ in R such that

$$\phi - \Sigma_j r_j^{(0)} \phi_j^{(0)} = a_1 X + o(1)$$

(say). Since ϕ and the $\phi_j^{(0)}$ are in $\mathfrak{A}$, $a_1 \in \mathfrak{a}_1$, and therefore we can find elements $r_j^{(1)} \in R$ such that

$$\phi - \Sigma_j r_j^{(0)} \phi_j^{(0)} - \Sigma_j r_j^{(1)} \phi_j^{(1)} = a_2 X^2 + o(2).$$

We next observe that $a_2 \in \mathfrak{a}_2$. Proceeding in this way, we obtain elements $r_j^{(i)}$ $(1 \leqslant j \leqslant s, 0 \leqslant i \leqslant m-1)$ such that

$$\left.\begin{aligned} \phi - \Sigma_j r_j^{(0)} \phi_j^{(0)} - \Sigma_j r_j^{(1)} \phi_j^{(1)} - \ldots - \Sigma_j r_j^{(m-1)} \phi_j^{(m-1)} \\ = a_m X^m + o(m) = \psi \text{ (say)} \end{aligned}\right\} \tag{A}$$

The proof will now be completed by showing that ψ is in the ideal $(\phi_1^{(m)}, \phi_2^{(m)}, \ldots, \phi_s^{(m)})$. From (A), we see that $a_m \in \mathfrak{a}_m$, hence we can find elements $r_j^{(m)} \in R$ such that

$$\psi - \Sigma_j r_j^{(m)} \phi_j^{(m)} = a_{m+1} X^{m+1} + o(m+1).$$

Then $a_{m+1} \in \mathfrak{a}_{m+1} = \mathfrak{a}_m$, hence, with suitable $r_j^{(m+1)} \in R$,

$$\psi - \Sigma_j r_j^{(m)} \phi_j^{(m)} - \Sigma_j r_j^{(m+1)} X \phi_j^{(m)} = a_{m+2} X^{m+2} + o(m+2).$$

We now determine elements $r_j^{(m+2)} \in R$ so that

$$\psi - \Sigma_j r_j^{(m)} \phi_j^{(m)} - \Sigma_j r_j^{(m+1)} X \phi_j^{(m)} - \Sigma_j r_j^{(m+2)} X^2 \phi_j^{(m)} = a_{m+3} X^{m+3} + o(m+3),$$

and so on. In this way we generate s sequences $r_j^{(m)}, r_j^{(m+1)}, r_j^{(m+2)}, \ldots$, where $1 \leqslant j \leqslant s$. Put

$$\chi_j = r_j^{(m)} + r_j^{(m+1)} X + r_j^{(m+2)} X^2 + \ldots,$$

then $\psi = \Sigma_j \chi_j \phi_j^{(m)}$. For

$$\begin{aligned} \psi - \Sigma_j \chi_j \phi_j^{(m)} &= \psi - \Sigma_j r_j^{(m)} \phi_j^{(m)} - \Sigma_j r_j^{(m+1)} X \phi_j^{(m)} - \ldots - \Sigma_j r_j^{(m+n)} X^n \phi_j^{(m)} \\ &\quad + \Sigma_j \phi_j^{(m)} (r_j^{(m)} + r_j^{(m+1)} X + \ldots + r_j^{(m+n)} X^n - \chi_j) \\ &= o(m+n) + o(m+n) = o(m+n). \end{aligned}$$

Since this holds for all n, it follows that $\psi - \Sigma_j \chi_j \phi_j^{(m)} = 0$. This completes the proof.

PROPOSITION 5. *Let Q be a local ring, then*

$$Q^* = Q[[X_1, X_2, \ldots, X_n]]$$

is also a local ring and

$$\mathfrak{m}^* = Q^*\mathfrak{m} + Q^*X_1 + Q^*X_2 + \ldots + Q^*X_n.$$

Proof. By Theorem 3, Q^* is Noetherian. Put

$$\mathfrak{m}^* = Q^*\mathfrak{m} + Q^*X_1 + \ldots + Q^*X_n,$$

then $1 \notin \mathfrak{m}^*$, so that $\mathfrak{m}^*$ is a proper ideal. Let $\phi \in Q^* - \mathfrak{m}^*$; the proof will be complete if we show that ϕ has an inverse in Q^*. Put $\phi = a + \phi_1$ where $a \in Q$ and where $\phi_1 \in (X_1, X_2, \ldots, X_n)$, then $\phi_1 \in \mathfrak{m}^*$, consequently, since $\phi \notin \mathfrak{m}^*$, $a \notin \mathfrak{m}$. Thus a is a unit in Q and therefore we may write $\phi = a(1-\psi)$, where $\psi \in (X_1, X_2, \ldots, X_n)$. It is now enough to show that $1-\psi$ has an inverse, and this we do as follows. If $(\nu) = (\nu_1, \nu_2, \ldots, \nu_n)$ is a set of non-negative integers, the coefficient of $X_1^{\nu_1} X_2^{\nu_2} \ldots X_n^{\nu_n}$ in $1 + \psi + \psi^2 + \ldots + \psi^m$ is the same for all large values of m. Denote this coefficient by $a_{(\nu)}$ and put $\chi = \sum_{(\nu)} a_{(\nu)} X_1^{\nu_1} X_2^{\nu_2} \ldots X_n^{\nu_n}$, then

$$\begin{aligned} \chi(1-\psi) - 1 &= (1 + \psi + \ldots + \psi^m)(1-\psi) - 1 + o(m) \\ &= -\psi^{m+1} + o(m) = o(m). \end{aligned}$$

But this holds for all m, accordingly $\chi(1-\psi) - 1 = 0$, so that χ is the required inverse.

5·5. The completion of a local ring. Let Q be a local ring. A local ring $\bar{Q}$ will be called *a completion of* Q if

(1) *$\bar{Q}$ is a concordant extension of Q,*

(2) *$\bar{Q}$ is complete,*

(3) *Every element of $\bar{Q}$ is the limit of a sequence of elements of Q.*

THEOREM 4. *Every local ring Q has a completion.*

Proof. Let $\mathfrak{m}=(u_1, u_2, \ldots, u_n)$, where, however, the base need not be minimal. Put $Q^*=Q[[X_1, X_2, \ldots, X_n]]$, then, by Proposition 5, Q^* is a local ring and $\mathfrak{m}^*=Q^*\mathfrak{m}+Q^*X_1+\ldots+Q^*X_n$. If ϕ is a power series, we shall use ϕ_k and $\phi_k(X)$ to denote the sum of the terms in ϕ which are of degree k; accordingly $\phi\equiv\phi_0+\phi_1+\ldots+\phi_k$ modulo $(\mathfrak{m}^*)^{k+1}$, and therefore $\phi_0+\phi_1+\ldots+\phi_k\to\phi$ in Q^* as $k\to\infty$. We now form a set $\mathfrak{n}$ of power series by putting a power series ϕ into $\mathfrak{n}$ if, and only if, $\phi_0(u)+\phi_1(u)+\ldots+\phi_k(u)\to 0$ in Q as $k\to\infty$. Here $\phi_m(u)$ is obtained from $\phi_m(X)$ by replacing $X_1, X_2, \ldots, X_n$ by $u_1, u_2, \ldots, u_n$ respectively. We assert that $\mathfrak{n}$ is an ideal. To see this let $\phi\in\mathfrak{n}$, let $\phi'\in\mathfrak{n}$, and let ψ be an arbitrary power series. It is evident that $\phi\pm\phi'\in\mathfrak{n}$, and that, if $\chi=\phi\psi$, then

$$\chi_k(u)=\phi_0(u)\,\psi_k(u)+\phi_1(u)\,\psi_{k-1}(u)+\ldots+\phi_k(u)\,\psi_0(u).$$

Let an integer $s\geqslant 0$ be given and choose m_0 so that

$$\phi_0(u)+\phi_1(u)+\ldots+\phi_m(u)\in\mathfrak{m}^s$$

if $m\geqslant m_0$. If now $m>m_0+s$ we have

$$\begin{aligned}\chi_0(u)&+\chi_1(u)+\ldots+\chi_m(u)\\ &\equiv\psi_0(u)\,[\phi_0(u)+\ldots+\phi_m(u)]+\psi_1(u)\,[\phi_0(u)+\ldots+\phi_{m-1}(u)]\\ &\qquad+\ldots+\psi_{m-m_0}(u)\,[\phi_0(u)+\ldots+\phi_{m_0}(u)]\pmod{\mathfrak{m}^s}\\ &\equiv 0\pmod{\mathfrak{m}^s}.\end{aligned}$$

This shows that $\chi\in\mathfrak{n}$, and thereby establishes that $\mathfrak{n}$ is an ideal, which, we note, is certainly proper. Put $\bar{Q}=Q^*/\mathfrak{n}$ so that $\bar{Q}$ is a local ring. From the definition of $\mathfrak{n}$ it follows at once that $\mathfrak{n}\cap Q=(0)$, and therefore, in the mapping of Q^* on to $\bar{Q}$, Q will be mapped isomorphically. We identify Q with its image in $\bar{Q}$ and, from now on, regard $\bar{Q}$ as an extension ring of Q. The definition of $\mathfrak{n}$ also shows that $X_i - u_i \in \mathfrak{n}$, hence $\bar{\mathfrak{m}} = \bar{Q}u_1 + \bar{Q}u_2 + \ldots + \bar{Q}u_n$, or, as we may write it,

$$\bar{\mathfrak{m}} = \bar{Q}\mathfrak{m}. \tag{5·5·1}$$

Next, let $a \in Q \cap \bar{\mathfrak{m}}^s$. Since $X_i - u_i \in \mathfrak{n}$, $\bar{\mathfrak{m}}^s$ can be written in the form $\bar{\mathfrak{m}}^s = [\mathfrak{n} + (X_1, X_2, \ldots, X_n)^s]/\mathfrak{n}$, and therefore we can find $\phi \in (X_1, X_2, \ldots, X_n)^s$ such that $a \equiv \phi (\text{mod } \mathfrak{n})$. By the choice of ϕ,

$$a - \phi_s(u) - \phi_{s+1}(u) - \ldots - \phi_{s+m}(u) \to 0$$

in Q as $m \to \infty$, and this shows† that $a \in \mathfrak{m}^s$. But a was an arbitrary element of $Q \cap \bar{\mathfrak{m}}^s$, consequently $Q \cap \bar{\mathfrak{m}}^s \subseteq \mathfrak{m}^s$, and as the opposite inclusion follows from (5·5·1) we have

$$\bar{\mathfrak{m}}^s \cap Q = \mathfrak{m}^s. \tag{5·5·2}$$

This, combined with (2) of Lemma 2, shows that $\bar{Q}$ is a concordant extension of Q.

Let $\bar{\alpha} \in \bar{Q}$ and choose a power series ϕ so that $\bar{\alpha}$ is the residue of ϕ modulo $\mathfrak{n}$. Since $\phi_0(X) + \phi_1(X) + \ldots + \phi_k(X) \to \phi$ in Q^*, $\phi_0(u) + \phi_1(u) + \ldots + \phi_k(u) \to \bar{\alpha}$ in $\bar{Q}$ (Proposition 4), which shows that $\bar{\alpha}$ is the limit, in $\bar{Q}$, of a sequence of elements of Q. All that now remains to be established, in order to complete the proof, is the completeness of $\bar{Q}$. Let $(\bar{A}_m)$ be a Cauchy sequence in $\bar{Q}$ and put $\bar{a}_m = \bar{A}_m - \bar{A}_{m-1}$ $(\bar{a}_0 = \bar{A}_0)$. By Lemma 1, $\bar{a}_m \to 0$ in $\bar{Q}$. Choose a sequence (s_m) of integers such that $s_m \to \infty$ as $m \to \infty$ and such that $\bar{a}_m \in \bar{\mathfrak{m}}^{s_m}$ for all m. We have, as has already been observed,

$$\bar{\mathfrak{m}}^s = [(X_1, X_2, \ldots, X_n)^s + \mathfrak{n}]/\mathfrak{n},$$

consequently we can find a power series $\phi^{(m)} \in (X_1, X_2, \ldots, X_n)^{s_m}$ such that $\bar{a}_m$ is the residue of $\phi^{(m)}$ modulo $\mathfrak{n}$. Since $s_m \to \infty$, the coefficient $c_{(\nu)}$ of $X_1^{\nu_1} X_2^{\nu_2} \ldots X_n^{\nu_n}$ in $\phi^{(0)} + \phi^{(1)} + \ldots + \phi^{(m)}$ is independent of m if m is sufficiently large. Put $\phi = \sum_{(\nu)} c_{(\nu)} X_1^{\nu_1} X_2^{\nu_2} \ldots X_n^{\nu_n}$ then, for all m,

$$\phi - \phi^{(0)} - \phi^{(1)} - \ldots - \phi^{(m)} \in (X_1, X_2, \ldots, X_n)^{t_m},$$

where $t_m = \min(s_{m+1}, s_{m+2}, \ldots)$. Also, $t_m \to \infty$ as $m \to \infty$ and $(X_1, X_2, \ldots, X_n)^{t_m} \subseteq (\mathfrak{m}^*)^{t_m}$, consequently $\phi^{(0)} + \phi^{(1)} + \ldots + \phi^{(m)} \to \phi$ in Q^*, and therefore, by Proposition 4, $\bar{a}_0 + \bar{a}_1 + \ldots + \bar{a}_m = \bar{A}_m$ tends to the residue of ϕ modulo $\mathfrak{n}$. This shows that $\bar{Q}$ is complete and establishes the theorem.

† The argument makes use of Proposition 2.

It will now be shown that if Q and Q' are isomorphic local rings then each completion of Q is isomorphic to each completion of Q'. In preparation for the proof of this result, we note that if σ is an isomorphism of Q on to Q' then (necessarily) $\sigma(\mathfrak{m}) = \mathfrak{m}'$ and therefore $\sigma(\mathfrak{m}^s) = (\mathfrak{m}')^s$ for every integer s. Let (a_n) and (a'_n) be corresponding sequences in Q and Q' respectively so that $a'_n = \sigma(a_n)$ for all n. The relation $\sigma(\mathfrak{m}^s) = (\mathfrak{m}')^s$ shows that

(i) *If one of the sequences is a Cauchy sequence so is the other*, and

(ii) *If one of the sequences has a limit so has the other and the limits correspond under* σ.

THEOREM 5. *Let σ be an isomorphism of a local ring Q on to a local ring Q', let $\bar{Q}$ be a completion of Q and let $\bar{Q}'$ be a completion of Q'. Then σ can be extended to an isomorphism of $\bar{Q}$ on to $\bar{Q}'$ and this can be done in only one way.*

Proof. Let $\xi \in \bar{Q}$. We can choose a sequence (a_n) of elements of Q such that $a_n \to \xi$ in $\bar{Q}$. Then (a_n) is a Cauchy sequence in $\bar{Q}$ and in Q; consequently, if $a'_n = \sigma(a_n)$, (a'_n) is a Cauchy sequence in Q' and therefore it has a limit in $\bar{Q}'$. Let us note that if (b_n) is also a sequence of elements of Q tending to ξ in $\bar{Q}$, then $a_n - b_n \to 0$ in Q and therefore $\sigma(a_n) - \sigma(b_n) \to 0$ in Q' and also in $\bar{Q}'$. Thus $\lim \sigma(a_n)$ and $\lim \sigma(b_n)$ both exist in $\bar{Q}'$ and they are equal; their common value we denote by $\phi(\xi)$. It is easily verified, and we leave the verification to the reader, that ϕ is a 1-1 mapping of $\bar{Q}$ on to $\bar{Q}'$. If $\xi \in Q$ we can take $a_n = \xi$ for all n, and this shows that $\phi(\xi) = \sigma(\xi)$; thus ϕ extends σ. Further, it follows from the continuity of addition and multiplication (Proposition 1) that if ξ_1 and ξ_2 are in $\bar{Q}$ then $\phi(\xi_1 + \xi_2) = \phi(\xi_1) + \phi(\xi_2)$. Our combined remarks now show that ϕ is an isomorphism of $\bar{Q}$ on to $\bar{Q}'$ extending σ.

Assume that ψ is also an isomorphism of $\bar{Q}$ on to $\bar{Q}'$ which extends σ. If now ξ and (a_n) are as before, then, since $a_n \to \xi$ in $\bar{Q}$, $\psi(a_n) \to \psi(\xi)$ in $\bar{Q}'$. But $a_n \in Q$, consequently $\psi(a_n) = \sigma(a_n)$ and so $\sigma(a_n) \to \psi(\xi)$. Thus $\psi(\xi) = \lim \sigma(a_n) = \phi(\xi)$, and, because this holds for all $\xi \in \bar{Q}$, $\psi = \phi$. This completes the proof.

Theorem 5 may be regarded as asserting that a local ring has essentially only one completion, so, from now on, we shall speak of *the* completion of a local ring. A bar will be used to indicate the completion of a given local ring; for example, if we are speaking of a local ring Q_0' then its completion will be denoted by $\bar{Q}_0'$.

PROPOSITION 6. *Let Q be a local ring and let $\bar{Q}$ be its completion, then $\bar{\mathfrak{m}} = \bar{Q}\mathfrak{m}$ and, for every integer $s \geqslant 0$, $\bar{\mathfrak{m}}^s \cap Q = \mathfrak{m}^s$. Further, if $\mathfrak{a}$ is an ideal in Q then $\bar{Q}\mathfrak{a} \cap Q = \mathfrak{a}$.*

Proof. The first two assertions follow from (5·5·1), (5·5·2) and the uniqueness of the completion. The last assertion will be established if we show that $\bar{Q}\mathfrak{a} \cap Q \subseteq \mathfrak{a}$. It will be proved that if $x \in \bar{Q}\mathfrak{a} \cap Q$ and $s > 0$ is an integer then $x \in \mathfrak{a} + \mathfrak{m}^s$, and from this the relation $\bar{Q}\mathfrak{a} \cap Q \subseteq \mathfrak{a}$ will follow at once because $\bigcap\limits_{s=1}^{\infty} (\mathfrak{a} + \mathfrak{m}^s) = \mathfrak{a}$. Since $x \in \bar{Q}\mathfrak{a}$, x can be written in the form

$$x = \bar{r}_1 a_1 + \bar{r}_2 a_2 + \ldots + \bar{r}_n a_n,$$

where $\bar{r}_i \in \bar{Q}$ and where $a_i \in \mathfrak{a}$. For each i choose $r_i \in Q$ so that $\bar{r}_i \equiv r_i (\bar{\mathfrak{m}}^s)$, and then put

$$r_1 a_1 + r_2 a_2 + \ldots + r_n a_n = a$$

so that $a \in \mathfrak{a}$. We now have

$$x - a = \Sigma(\bar{r}_i - r_i)\, a_i \in \bar{\mathfrak{m}}^s,$$

consequently $$x - a \in \bar{\mathfrak{m}}^s \cap Q = \mathfrak{m}^s$$

and therefore $x \in \mathfrak{a} + \mathfrak{m}^s$ as required.

Let us suppose that $\mathfrak{a}$ is a proper ideal of Q, then $\bar{Q}\mathfrak{a}$ is a proper ideal in $\bar{Q}$. The natural mapping of $\bar{Q}$ on to $\bar{Q}/\bar{Q}\mathfrak{a}$ will induce a homomorphism on Q which will have $\bar{Q}\mathfrak{a} \cap Q = \mathfrak{a}$ as its kernel, consequently, by the usual process of identification, we may say that $\bar{Q}/\bar{Q}\mathfrak{a}$ is an extension ring of $Q/\mathfrak{a}$. We use this remark in

THEOREM 6. *Let $\mathfrak{a}$ be a proper ideal in Q, then $\bar{Q}/\bar{Q}\mathfrak{a}$ is the completion of $Q/\mathfrak{a}$.*

Proof. By Proposition 4, $\bar{Q}/\bar{Q}\mathfrak{a} = \tilde{Q}_0$ (say) is complete; further, by the same proposition, since every element of $\bar{Q}$ is the limit of a sequence of elements of Q, every element of $\tilde{Q}_0$ is the limit of a sequence of elements of $Q/\mathfrak{a} = Q_0$ (say). It remains for us to show that Q_0 and $\tilde{Q}_0$ are concordant. Now $\bar{Q}\mathfrak{a} + \bar{\mathfrak{m}}^s = \bar{Q}(\mathfrak{a} + \mathfrak{m}^s)$, consequently, by Proposition 6, $(\bar{Q}\mathfrak{a} + \bar{\mathfrak{m}}^s) \cap Q = \mathfrak{a} + \mathfrak{m}^s$. This shows that in the residue rings $\tilde{Q}_0$ and Q_0 we have $\tilde{\mathfrak{m}}_0^s \cap Q_0 = \mathfrak{m}_0^s$, and now, by applying the second criterion of Lemma 2, it follows that Q_0 and $\tilde{Q}_0$ are concordant.

THEOREM 7. *Let Q be a local ring, let Q^* be a complete local ring which is an extension ring of Q, and let $\mathfrak{m}^* \cap Q = \mathfrak{m}$. Form the set Q' which consists of every element of Q^* that is a limit, in Q^*, of a sequence of elements of Q. Then Q' is a complete local ring; Q' and Q^* are concordant; and Q' is a homomorphic image of the completion $\bar{Q}$ of Q. Further, if Q and Q^* are concordant then Q' is the completion of Q.*

Proof. Let $\bar{Q}$ be a completion of Q, let $\xi \in \bar{Q}$ and let $a_n \to \xi$ in $\bar{Q}$ where $a_n \in Q$, then (a_n) is a Cauchy sequence in $\bar{Q}$ and in Q. Since $\mathfrak{m} \subseteq \mathfrak{m}^*$, for every integer s, $\mathfrak{m}^s \subseteq (\mathfrak{m}^*)^s$, and this shows that (a_n) is a Cauchy sequence in Q^*. By hypothesis, Q^* is complete, consequently (a_n) has a limit in Q^*. If this limit be denoted by $\phi(\xi)$ then (by a slight modification of arguments used in the proof of Theorem 5) it follows that, first, $\phi(\xi)$ depends only on ξ and not on the choice of the sequence (a_n); and, secondly, ϕ is a homomorphism of $\bar{Q}$ into Q^* which induces the identity mapping on Q. Let us denote the image of $\bar{Q}$, under the mapping ϕ, by Q_0', then, since $\phi(\xi)$ is the limit of a sequence of elements of Q, $\phi(\xi) \in Q'$ and therefore $Q_0' \subseteq Q'$. It will be shown that $Q_0' = Q'$. By Proposition 4, Q_0' is complete. Assume, for the moment, that $\xi \in \bar{\mathfrak{m}}$, then, if n is sufficiently large $a_n = (a_n - \xi) + \xi \in \bar{\mathfrak{m}}$ and therefore $a_n \in \bar{\mathfrak{m}} \cap Q = \mathfrak{m}$. It follows that, for all sufficiently large n, $a_n \in \mathfrak{m}^*$, consequently, by Proposition 2, $\phi(\xi) \in \mathfrak{m}^*$. But ξ was any element of $\bar{\mathfrak{m}}$, accordingly $\mathfrak{m}_0' = \phi(\bar{\mathfrak{m}}) \subseteq \mathfrak{m}^*$ and therefore $Q_0' \cap \mathfrak{m}^* = \mathfrak{m}_0'$. Combining this with Theorem 2 we see that Q_0' and Q^* are concordant.

Let $\alpha' \in Q'$ and let $b_n \to \alpha'$ in Q^* where $b_n \in Q$. The sequence (b_n) is a Cauchy sequence in Q^*, consequently, since Q_0' and Q^* are concordant, it is a Cauchy sequence in Q_0' and therefore it has a limit α_0' in Q_0'. Thus $b_n \to \alpha_0'$ in Q_0', hence, because Q_0' and Q^* are concordant, $b_n \to \alpha_0'$ in Q^*. This shows that $\alpha' = \alpha_0' \in Q_0'$ and thereby establishes that $Q' = Q_0'$. All the assertions of the theorem are now proved except the last. Assume that Q and Q^* are concordant. We shall prove that Q' is a completion of Q by showing that the homomorphism ϕ is, in fact, an isomorphism. Let ξ and (a_n) be as at first (i.e. ξ is no longer assumed to be in $\bar{\mathfrak{m}}$) and suppose that $\phi(\xi)=0$. Then $a_n \to 0$ in Q^* and therefore, since Q and Q^* are concordant, $a_n \to 0$ in Q. This shows that $a_n \to 0$ in $\bar{Q}$ so that $\xi=0$. Thus ϕ is an isomorphism and the proof is complete.

5·6. Some transition theorems. By a Transition Theorem we mean a result telling us how a property, which holds in Q, will be affected when we go over to the completion $\bar{Q}$.

LEMMA 3. *Let $\bar{\mathfrak{q}}=(\bar{a}_1, \bar{a}_2, \ldots, \bar{a}_s)$ be an $\bar{\mathfrak{m}}$-primary ideal in the completion $\bar{Q}$ of Q, then we can find s elements $a_1, a_2, \ldots, a_s$ in Q such that $\mathfrak{q}=Qa_1+Qa_2+\ldots+Qa_s$ is $\mathfrak{m}$-primary and $\bar{\mathfrak{q}}=\bar{Q}\mathfrak{q}$.*

Proof. Since $\bar{\mathfrak{q}}$ is $\bar{\mathfrak{m}}$-primary, we can find an integer n such that $\bar{\mathfrak{m}}^n \subseteq \bar{\mathfrak{q}}$, and then, since $\bar{Q}$ is the completion of Q, we can find elements $a_1, a_2, \ldots, a_s$ in Q with the property that $\bar{a}_i \equiv a_i(\bar{\mathfrak{m}}^{n+1})$ for $1 \leqslant i \leqslant s$. Put $\mathfrak{q}=Qa_1+Qa_2+\ldots+Qa_s$, then from $\bar{a}_i \equiv a_i(\bar{\mathfrak{m}}^{n+1})$ we obtain $\bar{a}_i \equiv a_i(\bar{\mathfrak{q}})$, which shows that $a_i \in \bar{\mathfrak{q}}$ and that $\bar{Q}\mathfrak{q} \subseteq \bar{\mathfrak{q}}$. Again, if $\bar{x} \in \bar{\mathfrak{q}}$ we can write $\bar{x}$ in the form $\bar{x}=\bar{r}_1\bar{a}_1+\bar{r}_2\bar{a}_2+\ldots+\bar{r}_s\bar{a}_s$, where $\bar{r}_i \in \bar{Q}$, consequently $\bar{x} \equiv \bar{r}_1 a_1+\bar{r}_2 a_2+\ldots+\bar{r}_s a_s(\bar{\mathfrak{m}}^{n+1})$ and therefore $\bar{x} \in \bar{Q}\mathfrak{q}+\bar{\mathfrak{m}}^{n+1} \subseteq \bar{Q}\mathfrak{q}+\bar{\mathfrak{m}}\bar{\mathfrak{q}}$. We now see that $\bar{\mathfrak{q}} \subseteq \bar{Q}\mathfrak{q}+\bar{\mathfrak{m}}\bar{\mathfrak{q}}$, hence $\bar{\mathfrak{q}} \subseteq \bar{Q}\mathfrak{q}$ (§4·2, Proposition 1) and therefore $\bar{\mathfrak{q}}=\bar{Q}\mathfrak{q}$. Further, $\bar{\mathfrak{q}} \cap Q=\bar{Q}\mathfrak{q} \cap Q=\mathfrak{q}$ (Proposition 6) and therefore, since $\bar{\mathfrak{q}}$ is $\bar{\mathfrak{m}}$-primary, it follows, by Proposition 8 of §2·4, that $\mathfrak{q}$ is a primary ideal belonging to $\bar{\mathfrak{m}} \cap Q=\mathfrak{m}$. The proof is now complete.

PROPOSITION 7. *There is a 1-1 correspondence between the $\mathfrak{m}$-primary ideals $\mathfrak{q}$ in Q and the $\bar{\mathfrak{m}}$-primary ideals $\bar{\mathfrak{q}}$ in $\bar{Q}$ such that if $\mathfrak{q}$ and $\bar{\mathfrak{q}}$ correspond then $\bar{\mathfrak{q}}=\bar{Q}\mathfrak{q}$ and $\mathfrak{q}=\bar{\mathfrak{q}} \cap Q$.*

Proof. Let $\mathfrak{q}$ be an $\mathfrak{m}$-primary ideal and put $\bar{\mathfrak{q}}=\bar{Q}\mathfrak{q}$. We can find an integer n such that $\mathfrak{m}^n \subseteq \mathfrak{q}$ and then

$$\bar{\mathfrak{m}}^n=\bar{Q}\mathfrak{m}^n \subseteq \bar{Q}\mathfrak{q}=\bar{\mathfrak{q}},$$

which shows that $\bar{\mathfrak{q}}$ is $\bar{\mathfrak{m}}$-primary. Moreover, by Proposition 6, $\bar{\mathfrak{q}} \cap Q=\bar{Q}\mathfrak{q} \cap Q=\mathfrak{q}$, and now the proposition will be established if we show that every $\bar{\mathfrak{m}}$-primary ideal is the extension of an $\mathfrak{m}$-primary ideal. This, however, is shown by Lemma 3.

THEOREM 8. *If $\bar{Q}$ is the completion of Q then* $\dim \bar{Q} = \dim Q$. *Further, a set $(b_1, b_2, \ldots, b_s)$ of elements of Q is a system of parameters in Q if, and only if, it is a system of parameters in $\bar{Q}$.*

Proof. Let $\dim Q = d$ and let $\dim \bar{Q} = p$. In order to dispose of the case in which one of d and p is zero, we note that a local ring is of dimension zero when, and only when, it is a primary ring, and this will occur when, and only when, some power of the maximal ideal is zero. But $\bar{\mathfrak{m}}^s = \bar{Q}\mathfrak{m}^s$ and $\mathfrak{m}^s = Q \cap \bar{\mathfrak{m}}^s$ hence $\mathfrak{m}^s = (0)$ if, and only if, $\bar{\mathfrak{m}}^s = (0)$, and therefore $d = 0$ if, and only if, $p = 0$. It will now be supposed that $d \geqslant 1$ and $p \geqslant 1$. By Theorem 1 of §4·2, it is possible to find d elements which will generate an $\mathfrak{m}$-primary ideal, consequently, since the same d elements will generate in $\bar{Q}$ an $\bar{\mathfrak{m}}$-primary ideal, $p \leqslant d$. It is also possible to find p elements (in $\bar{Q}$) which will generate an $\bar{\mathfrak{m}}$-primary ideal, and now it follows from Lemma 3 that an $\mathfrak{m}$-primary ideal can also be generated by p elements. This shows that $d \leqslant p$ and so the first assertion has been proved. To establish the second assertion, we have only to combine $\dim Q = \dim \bar{Q}$ with the remark that $b_1, b_2, \ldots, b_s$ will generate an $\mathfrak{m}$-primary ideal in Q, if, and only if, they generate an $\bar{\mathfrak{m}}$-primary ideal in $\bar{Q}$.

COROLLARY. *A local ring Q is regular when, and only when, its completion $\bar{Q}$ is regular.*

Proof. Let $(u_1, u_2, \ldots, u_n)$ be a *minimal* base of $\mathfrak{m}$ then, since $\bar{\mathfrak{m}} = \bar{Q}\mathfrak{m}$, it is also a base of $\bar{\mathfrak{m}}$. We assert that, as a base of $\bar{\mathfrak{m}}$, $(u_1, u_2, \ldots, u_n)$ is minimal. For suppose, for example, that

$$\bar{\mathfrak{m}} = \bar{Q}u_2 + \ldots + \bar{Q}u_n = \bar{Q}(Qu_2 + \ldots + Qu_n),$$

then, by Proposition 6, $\mathfrak{m} = \bar{\mathfrak{m}} \cap Q = Qu_2 + \ldots + Qu_n$, and this contradicts our hypothesis. By the definition of regularity, Q is regular if, and only if, $n = \dim Q$, while $\bar{Q}$ is regular if, and only if, $n = \dim \bar{Q}$. Since $\dim Q = \dim \bar{Q}$, this establishes the corollary.

PROPOSITION 8. *Let $\bar{Q}$ be the completion of Q and let $t_1, t_2, \ldots, t_p$ be elements of Q; then the t_i are analytically independent in Q if, and only if, they are analytically independent in $\bar{Q}$.*

Proof. It follows immediately, from $\bar{\mathfrak{m}} \cap Q = \mathfrak{m}$ and from the definitions, that if the t_i are analytically independent in $\bar{Q}$ then they must be analytically independent in Q. We assume, therefore, that the t_i are analytically independent in Q and that $\bar{\phi}(t_1, t_2, \ldots, t_p) = 0$, where $\bar{\phi}$ is a form of degree s with coefficients in $\bar{Q}$. We wish to prove that the coefficients of $\bar{\phi}$ are in $\bar{\mathfrak{m}}$. Choose a form ϕ of degree s so that ϕ has its coefficients in Q and so that all the coefficients of $\bar{\phi} - \phi$ are in $\bar{\mathfrak{m}}$, then it will suffice to prove that the coefficients of ϕ are in $\mathfrak{m}$. Now

$$\phi(t) = \phi(t_1, t_2, \ldots, t_p) = \phi(t) - \bar{\phi}(t) \equiv 0(\bar{Q}\mathfrak{a}),$$

where $\mathfrak{a} = \mathfrak{m}(t_1, t_2, \ldots, t_p)^s$, consequently

$$\phi(t) \in \bar{Q}\mathfrak{a} \cap Q = \mathfrak{a} = \mathfrak{m}(t_1, t_2, \ldots, t_p)^s.$$

But, as was observed in §4·4, since the t_i are analytically independent in Q, $\phi(t)$ cannot belong to $\mathfrak{m}(t_1, t_2, \ldots, t_p)^s$ unless all the coefficients of ϕ are in $\mathfrak{m}$. This completes the proof.

THEOREM 9. *Let $\bar{Q}$ be the completion of Q, let $b \in Q$ and let $\mathfrak{a}$ be an ideal of Q; then $\bar{Q}[\mathfrak{a} : (b)] = \bar{Q}\mathfrak{a} : \bar{Q}b$.*

Proof. Let $\mathfrak{c} = \mathfrak{a} : (b)$ then $b\mathfrak{c} \subseteq \mathfrak{a}$ and therefore

$$\bar{Q}b\mathfrak{c} = (\bar{Q}b)\,(\bar{Q}\mathfrak{c}) \subseteq \bar{Q}\mathfrak{a},$$

which shows that $\bar{Q}[\mathfrak{a} : (b)] \subseteq \bar{Q}\mathfrak{a} : \bar{Q}b$. Now let $\bar{x} \in \bar{Q}\mathfrak{a} : \bar{Q}b$ and let $s \geqslant 0$ be an integer. Choose $x \in Q$ so that $\bar{x} \equiv x(\bar{\mathfrak{m}}^s)$, then

$$bx = b(x - \bar{x}) + b\bar{x} \in b\bar{\mathfrak{m}}^s + \bar{Q}\mathfrak{a}$$

and therefore

$$bx \in \bar{Q}[\mathfrak{a} + b\mathfrak{m}^s] \cap Q = \mathfrak{a} + b\mathfrak{m}^s.$$

This shows that, with a suitable $t \in \mathfrak{m}^s$, $b(x - t) \in \mathfrak{a}$, or, $x - t \in \mathfrak{a} : (b)$. Thus

$$x \in \mathfrak{a} : (b) + \mathfrak{m}^s \subseteq \bar{Q}[\mathfrak{a} : (b)] + \bar{\mathfrak{m}}^s,$$

and hence $\bar{x} \in \bar{Q}[\mathfrak{a} : (b)] + \bar{\mathfrak{m}}^s$. But $\bar{x}$ was any element of $\bar{Q}\mathfrak{a} : \bar{Q}b$, consequently $\bar{Q}\mathfrak{a} : \bar{Q}b \subseteq \bar{Q}[\mathfrak{a} : (b)] + \bar{\mathfrak{m}}^s$ and this holds for all s. It follows that $\bar{Q}\mathfrak{a} : \bar{Q}b \subseteq \bar{Q}[\mathfrak{a} : (b)]$ and this completes the proof.

COROLLARY 1. *If c is not a zero divisor in Q then it is not a zero divisor in $\bar{Q}$.*

Proof. If c is not a zero divisor in Q then $(0) : (c) = (0)$. The theorem now shows that, in $\bar{Q}$, $(0) : \bar{Q}c = (0)$ and this establishes the corollary.

COROLLARY 2. *If $\mathfrak{a}$ and $\mathfrak{b}$ are ideals of Q and $\mathfrak{a} : \mathfrak{b} = \mathfrak{a}$, then $\bar{Q}\mathfrak{a} : \bar{Q}\mathfrak{b} = \bar{Q}\mathfrak{a}$.*

Proof. We may suppose that $\mathfrak{a}$ is a proper ideal and then, since $\mathfrak{a} : \mathfrak{b} = \mathfrak{a}$, $\mathfrak{b}$ is not contained in any of the prime ideals $\mathfrak{p}_1, \mathfrak{p}_2, \ldots, \mathfrak{p}_r$ (say) which belong to $\mathfrak{a}$ (see § 1·9, Theorem 6). By Proposition 6 of § 1·5, we can choose $b \in \mathfrak{b}$ so that b is not contained in any $\mathfrak{p}_i$, and when this has been done we shall have $\mathfrak{a} : (b) = \mathfrak{a}$. It follows from our theorem that $\bar{Q}\mathfrak{a} : \bar{Q}b = \bar{Q}\mathfrak{a}$, so that b, and *a fortiori* $\bar{Q}\mathfrak{b}$, is not contained in any prime ideal of $\bar{Q}\mathfrak{a}$, consequently $\bar{Q}\mathfrak{a} : \bar{Q}\mathfrak{b} = \bar{Q}\mathfrak{a}$ as required.

NOTES

In presenting a part of the theory of ideals in Noetherian rings, the author has tried to select topics which will suggest a continuous and natural development of his subject in a single main direction. To make this suggestion more effective, he has deliberately refrained from stating in the text what have been the individual contributions of the different writers on this branch of mathematics. The notes which follow will acquaint the reader with some of the important original papers, which mark the stages in the evolution of our subject. It is from these papers that the author has, very largely, constructed his account, and they are the sources to which the reader should turn if he wishes to extend his knowledge. For convenience, a list of articles and books to which reference is made has been printed on p. 110, and numbers in bold type occurring in these notes will refer to that list. No attempt at constructing anything like a full bibliography was ever considered; there already exists a very detailed account, compiled by W. Krull, of what had been done in ideal theory prior to 1935 (see (**8**)).

CHAPTER I

Rings, in which the maximal condition for ideals is satisfied, are now usually called 'Noetherian' after E. Noether, who first realized their importance. It was in 1921 that her now famous paper appeared (see (**11**)), in which all the main results on the existence and partial uniqueness of normal decompositions were proved. The treatment given in Chapter I is essentially that of Noether herself with one important difference. Noether considers isolated components of ideals and proves that they are uniquely determined by the corresponding prime ideals, but this is done without using the concept of the component of an ideal determined by a multiplicatively closed set. This fundamental and fruitful concept appears rather unobtrusively in a paper by B. L. van der Waerden on algebraic geometry, which was published in 1928 (15), and we have taken every opportunity, in this tract, to stress the usefulness of this notion.

The abstract theory of ideals was, of course, preceded by investigations of a more special nature. Previous to the publication of Noether's paper, the polynomial ring $C[X_1, X_2, \ldots, X_n]$, C being the field of complex numbers, had been studied in considerable detail by methods (such as elimination theory) which depended on the special nature of the ring. Among the names of those who were associated with this earlier period of our subject are D. Hilbert, L. Kronecker, E. Lasker and F. S. Macaulay. Still considering $C[X_1, X_2, \ldots, X_n]$, it may be of interest to recall that the Basis Theorem (asserting that every ideal is finitely generated) was proved by Hilbert in 1890, and that the main facts concerning the Primary Decomposition were established by Lasker in 1905. This work on polynomial rings not only served as a concrete model for the abstract theory and provided a rich supply of illustrative examples, but it formed a link connecting the general theory of Noetherian rings with the foundations of geometry; and as the abstract theory has grown to its present stature so, at the same time, the work on the basic algebraic problems of geometry has progressed with it. Lack of space prevents our examining, even superficially, the way in which ideal theory and algebraic geometry have stimulated each other, but, in this connexion, the reader will find it interesting and profitable to consult the tract by Macaulay (10), which is published in this series. Some of the most striking examples of recent applications of ideal theory to algebraic geometry will be found in the papers of C. Chevalley and O. Zariski.

It may perhaps be a matter for surprise that, so far, there has been no mention of the work of R. Dedekind. The modern theory of ideals has a double origin, and this has led to a branching of the subject. On one side we have the so-called multiplicative or classical theory, which has grown out of the theory of algebraic numbers, and on the other side we have the so-called additive theory, which has arisen out of the study of polynomial rings and certain aspects of algebraic geometry. When choosing the material to put in this tract, the author turned his attention almost exclusively to the additive theory, and in so doing he realizes that he has suppressed a rich and fascinating part of his subject. His main excuses are shortage of space and the fact that the multiplicative theory is, on the whole, fairly well known. (A short account will be found under the heading 'Axiomatische Begründung der klassichen Idealtheorie' in van der Waerden's well-known text-book (17)). For the beginner, who wishes to have a rough idea of the relation of the multiplicative to the additive theory, we may put the matter briefly, if very inadequately, as follows. The multiplicative theory can be regarded as a reformulation and extension of the additive theory which is made possible either by restricting one's attention to a very special kind of ring, or else by considering more general rings and treating as irrelevant all prime ideals whose rank is greater than unity. For example, the theory of divisors given in Chapter IV may be considered as forming a part of the multiplicative theory.

CHAPTER II

The remarks on homomorphisms and isomorphisms, like the 'Preliminaries', are intended solely for readers who have no previous acquaintance with the abstract theory of rings, for whom the author wished to make his account self-contained. Turning to the more sophisticated matter of the extensions and contractions of ideals, a very general study of these operations was made by H. Grell in 1927 (5); and, in particular, his account contains a discussion of the special case in which the extension ring is an ordinary ring of quotients of the subring. The concept of a generalized ring of quotients is due to C. Chevalley (2), and here it is interesting to note that he appears to have been led to this idea more because of its usefulness in applications to algebraic geometry, than because of the important role it can be made to play in the abstract theory. Chevalley confined his attention to Noetherian rings, and his argument, as it stands, is not applicable to general commutative rings. This was observed and remedied by I. Uzkov (13). Uzkov's treatment differs in matters of detail from that given here, which more closely resembles the one given originally by Chevalley.

It was remarked in the text that the name 'Local Ring' for a Noetherian ring which has only one maximal prime ideal is taken from geometry. We shall now indicate, very briefly, what is behind this remark. Suppose that P is a point on an irreducible variety V, which is situated in n-dimensional (affine) space. Two rational functions

$$f(x_1, x_2, \ldots, x_n)/g(x_1, x_2, \ldots, x_n)$$

and
$$\phi(x_1, x_2, \ldots, x_n)/\psi(x_1, x_2, \ldots, x_n)$$

—where it is supposed that neither g nor ψ vanishes identically on V—are called *equivalent on* V if

$$f(x_1, x_2, \ldots, x_n)\, \psi(x_1, x_2, \ldots, x_n) = \phi(x_1, x_2, \ldots, x_n)\, g(x_1, x_2, \ldots, x_n)$$

at every point of V. If now classes of rational functions, which are mutually equivalent on V, are formed then each class is called a *rational function on* V. Let ξ be a rational function on V and suppose that the coordinates of P are $(\alpha_1, \alpha_2, \ldots, \alpha_n)$, then ξ is said to be *finite at* P if ξ has a representative

$$f(x_1, x_2, \ldots, x_n)/g(x_1, x_2, \ldots, x_n)$$

such that $g(\alpha_1, \alpha_2, \ldots, \alpha_n)$ is not zero, and in this case we write

$$\xi(P) = \frac{f(\alpha_1, \alpha_2, \ldots, \alpha_n)}{g(\alpha_1, \alpha_2, \ldots, \alpha_n)},$$

and call $\xi(P)$ the *value of* ξ *at* P. It is easily seen that the functions ξ which are finite at P form a ring, if addition and multiplication are defined by means of representatives, and this ring is called the *local ring* $Q(P/V)$ *of* V *at* P. This ring is a local ring in the sense of the general theory; in fact, if ξ belongs to $Q(P/V)$ then ξ is a non-unit or a unit in $Q(P/V)$ according as $\xi(P)$ is or is not zero. Further (generally speaking), it turns out that geometric properties possessed by V in the neighbourhood of P correspond to algebraic properties of the ring $Q(P/V)$.

CHAPTER III

We owe the Intersection Theorem to Krull. It appears in a paper published in 1928 (7), where it is stated in a form which corresponds to our Theorem 4 of §3·2. Let us, however, consider another form of this result, namely, Theorem 2 of §3·1. If we have an ideal $\mathfrak{a}$ in a ring R we can obtain from $\mathfrak{a}$ a topology of the ring by regarding, for each element b, the sets $b+\mathfrak{a}$, $b+\mathfrak{a}^2$, $b+\mathfrak{a}^3, \ldots$ as forming a fundamental system of neighbourhoods of b. If we want the topology to be 'separated', that is, if we want different elements to have non-intersecting neighbourhoods, we find easily that this occurs when, and only when, the only element contained in all the powers of $\mathfrak{a}$ is the zero element. Theorem 2 therefore gives us a means of telling whether the topology determined by an ideal is of the separated type or not.

Every ring R is an Abelian group with respect to the rule of addition. This group, moreover, can be regarded as having the elements of R as operators; in fact, if r and a belong to R, and if we propose to regard r as an operator, then the result of operating with r on a will simply be the product ra. The theory of groups with operators (operator-groups) is therefore applicable to the theory of rings, and, as is clear from the definitions, in translating the group theory into the terminology of ideal theory operator-subgroups correspond exactly to ideals. From this point of view, our discussion of series of ideals in a primary ring is a special case of the theory of Jordan composition series for operator-groups. (Further applications of group theory to primary rings will be found in (6).) A paper published by van der Waerden in 1928 contains a section in which it is shown that the theory of composition series in a primary ring can be used to attach definite lengths to primary ideals, but, concerning that particular section, van der Waerden says that his discussion is a partial elaboration of some ideas of E. Noether (see (14), §3).

The basic results on the rank of a prime ideal in a Noetherian ring (see Theorems 6 and 7 of §3·5) date from 1928 (Krull (7)), and they mark a turning point in the development of the general theory. Previously, a Noetherian ring had been a kind of pale shadow of a polynomial ring, but after the publication of Krull's results the way was open for the introduction of a surprising amount of interesting detail. For example, it has been shown very recently that the Hilbert characteristic function of an ideal can be defined in a general local ring. This particular development was initiated by P. Samuel (**12**).

CHAPTERS IV AND V

Nearly the whole account of the theory of local rings has been based on three well-known papers, which, in chronological order, are by W. Krull (**9**), C. Chevalley (**1**) and I. S. Cohen (**4**) respectively. To some extent these overlap. The contents of Chapters IV–V should be regarded merely as an introduction to these papers (and to their applications to geometry), because the latter contain a great deal more than it has proved convenient to put into our account. In the brief remarks on the individual papers, which will be found below, mention is made of a few specially interesting results among those for which room could not be found in these pages.

In (**9**), Krull not only laid the foundations of the theory as it exists at present, but he also posed the problem of determining the structure of all complete local rings, and it is the solution of this problem which occupies a considerable part of Cohen's paper. Krull also establishes two theorems concerning regular local rings which are of unusual interest; the first states that a regular local ring is always integrally closed in its quotient field, while the second shows that (in a regular local ring) the sum of the rank and dimension of each prime ideal is equal to the dimension of the ring. One of the most important concepts introduced in (**9**) is that of the *ideal of leading forms* of a given ideal, and it is a little surprising to find that both Chevalley and Cohen make hardly any use of this concept, which surely has an important part to play in some of the future developments. Incidentally, (**9**) contains an example which shows that, in a Noetherian ring, an ideal need not be of finite dimension, although, as we know, it is necessarily of finite rank.

Let us, for the purpose of these notes, describe the local ring of an irreducible variety, at one of its points, as a *geometric local ring*, then geometric local rings have certain special properties not possessed by all local rings. It is for this reason that, in (1), Chevalley uses the term local ring in a more restricted sense than we have done, but his preliminary account of *semi-local rings* forms a natural and important extension of the general theory. A semi-local ring is a Noetherian ring which has only a *finite* number of maximal ideals and, considered as an extension of the notion of a local ring, the generalization is important for the following reason: a finite integral extension of a semi-local ring is again a semi-local ring, but a finite integral extension of a local ring (while it is necessarily semi-local) need not be a local ring. (In geometry, semi-local rings arise, for instance, when one wants to separate, by means of a birational transformation, the different analytic branches which a given variety has at one of its points.) A semi-local ring has a natural topology, namely, that determined by the ideal which is the intersection or (what is the same thing) the product of the maximal ideals. We can construct the completion of the ring with respect to this topology and much of the analytic theory of local rings carries over. In particular, the completion of a semi-local ring is again a semi-local ring. There is, in this connexion, a point which deserves a special mention. The completion of a semi-local ring is not only a semi-local ring, but also a *direct sum of local rings*. The components in the direct sum arise in this way. If we form the generalized ring of quotients of the uncompleted ring with respect to one of its maximal ideals, we obtain, by taking each of the maximal ideals in turn, a finite number of ordinary local rings. It is the completions of these ordinary local rings which are the components of the direct sum. Here we have another illustration of the usefulness of generalized rings of quotients (see (2), Prop. 5, p. 96).

As already remarked, Cohen (4) solved the problem of determining the structure of all complete local rings. For the case in which the complete ring Q has the same characteristic as its residue field P (this is the case which is important for geometry), Cohen's result is that Q is a homomorphic image of the power series ring $P[[X_1, X_2, \ldots, X_n]]$, where n is the number of elements in a minimal base of the maximal ideal of Q. Besides the fundamental structure theorems, Cohen proves many other interesting results of which we select two, namely,

(1) *If K is a field then the power series ring* $K[[X_1, X_2, \ldots, X_n]]$ *is a unique factorization domain*, and

(2) *In a regular local ring, an ideal which is of rank r and which can be generated by r elements is unmixed: that is to say all its prime ideals have rank r.*

The second of these theorems generalizes one given by Macaulay.

Krull, Chevalley and Cohen handle the problem of showing that every local ring has a completion which is also a local ring, in rather different ways, and the reader will find it interesting to compare the various methods used. The procedure followed in the text is, essentially, that of Chevalley.

Let us now consider, for a moment, geometric local rings. These can be more easily studied than abstract local rings and so we know more about them. (The extra knowledge that we have is, at the time of writing, largely duė to the work of Chevalley and Zariski.) We give three examples to illustrate the kind of results that have been obtained:

(*a*) *A regular geometric local ring is a unique factorization domain* (Zariski (18), Th. 5, p. 22).

(In connexion with this result, it should be noted that in Zariski's system of geometry a simple point on an irreducible variety is completely characterized by the property that its local ring is regular. This has added considerably to the interest taken in the study of regular local rings.)

(*b*) *The zero ideal of the completion of a geometric local ring is an intersection of prime ideals* (Chevalley (3), Th. 1, p. 11).

(*c*) *The completion of an integrally closed geometric local ring has no zero divisors* (Zariski (19)).

In conclusion, we recall that §§4·9 and 4·10, which deal with divisors, can be thought of as belonging to multiplicative ideal theory. The results of these sections constitute a variation of the well-known theory of van der Waerden on the subject of 'quasi-equal' ideals in an integrally closed Noetherian domain. The exposition given in the text is not the most simple, but it has, for us, the advantage of showing the relation of the theory of divisors to the more general theory. A particularly elegant and direct discussion, due to E. Artin, of the concept of 'quasi-equality' will be found in ((**17**), § 103), and there is an important application of these ideas in ((**16**), § 8).

REFERENCES

(1) Chevalley, C. On the theory of local rings. *Ann. Math., Princeton*, 44 (1943), 690–708.

(2) Chevalley, C. On the notion of the ring of quotients of a prime ideal. *Bull. Amer. Math. Soc.* 50 (1944), 93–97.

(3) Chevalley, C. Intersections of algebraic and algebroid varieties. *Trans. Amer. Math. Soc.* 57 (1945), 1–85.

(4) Cohen, I. S. On the structure and ideal theory of complete local rings. *Trans. Amer. Math. Soc.* 59 (1946), 54–106.

(5) Grell, H. Beziehungen zwischen den Idealen verschiedener Ringe. *Math. Ann.* 97 (1927), 490–523.

(6) Krull, W. Theorie und Anwendung der verallgemeinerten Abelschen Gruppen. *S.B. heidelberg. Akad. Wiss.* (1926), 1. Abh.

(7) Krull, W. Primidealketten in allgemeinen Ringbereichen. *S.B. heidelberg. Akad. Wiss.* (1928), 7. Abh.

(8) Krull, W. *Idealtheorie.* Ergebn. Math. iv, 3. New York (1948).

(9) Krull, W. Dimensionstheorie in Stellenringen. *J. reine angew. Math.* 179 (1938), 204–26.

(10) Macaulay, F. S. *Algebraic Theory of Modular Systems.* Cambridge tract 19 (1916).

(11) Noether, E. Idealtheorie in Ringbereichen. *Math. Ann.* 83 (1921), 24–66.

(12) Samuel, P. La notion de multiplicité en algèbre et en géometrie. Thesis (Paris), 1951.

(13) Uzkov, I. *On Rings of Quotients of Commutative Rings.* Amer. Math. Soc., Translations (Number 3) (1949).

(14) van der Waerden, B. L. On Hilbert's function, series of composition of ideals, and a generalization of a theorem of Bézout. *Proc. K. Acad. Wet. Amst.* 31 (1928), 749–70.

(15) van der Waerden, B. L. Eine Verallgemeinerung des Bézoutschen Theorems. *Math. Ann.* 99 (1928), 497–541.

(16) van der Waerden, B. L. Zur Produktzerlegung der Ideale in ganz abgeschlossenen Ringe. *Math. Ann.* 101 (1929), 293–308.

(17) van der Waerden, B. L. *Moderne Algebra*, 2nd ed. (1937). Springer.

(18) Zariski, O. The concept of a simple point of an abstract algebraic variety. *Trans. Amer. Math. Soc.* 62 (1947), 1–52.

(19) Zariski, O. Analytical irreducibility of normal varieties. *Ann. Math., Princeton*, 49 (1948), 352–61.

INDEX OF DEFINITIONS

The numbers refer to the pages on which the definitions will be found

Printed in the United States
By Bookmasters